P9-DNZ-302
2nd Edition

BASIC MATH & PRE-ALGEBRA SUPER REVIEW®

Staff of
Research & Education Association

Research & Education Association
Visit our website at: www.rea.com

Research & Education Association
61 Ethel Road West
Piscataway, New Jersey 08854
E-mail: info@rea.com

BASIC MATH & PRE-ALGEBRA SUPER REVIEW®

Printed in the United States of America

Library of Congress Control Number 2013932757

ISBN-13: 978-0-7386-1119-8
ISBN-10: 0-7386-1119-0

REA's *Basic Math & Pre-Algebra Super Review*®

Need help with Basic Math and Pre-Algebra? Want a quick review or refresher for class? This is the book for you!

REA's *Basic Math & Pre-Algebra Super Review*® gives you everything you need to know!

This *Super Review*® can be used as a supplement to your high school or college textbook, or as a handy guide for anyone who needs a fast review of the subject.

- **Comprehensive, yet concise coverage** – review covers the material that is typically taught in a beginning-level math and pre-algebra course. Each topic is presented in a clear and easy-to-understand format that makes learning easier.

- **Packed with practice** – each review lesson is packed with practice questions and answers for each topic. Practice what you've learned and build your basic math and pre-algebra skills, so you'll be ready for any problem you encounter on your next quiz or test.

- **Detailed answers** – our practice problems come with step-by-step detailed solutions to help you understand the material and sharpen your skills.

Whether you need a quick refresher on the subject, or are prepping for your next test, we think you'll agree that REA's *Super Review*® provides all you need to know!

Available Super Review® Titles

ARTS/HUMANITIES

Basic Music
Classical Mythology
History of Architecture
History of Greek Art

BUSINESS

Accounting
Macroeconomics
Microeconomics

COMPUTER SCIENCE

C++
Java

HISTORY

Canadian History
European History
United States History

LANGUAGES

English
French
French Verbs
Italian
Japanese for Beginners
Japanese Verbs
Latin
Spanish

MATHEMATICS

Algebra & Trigonometry
Basic Math & Pre-Algebra
Calculus
Geometry
Linear Algebra
Pre-Calculus
Statistics

SCIENCES

Anatomy & Physiology
Biology
Chemistry
Entomology
Geology
Microbiology
Organic Chemistry I & II
Physics

SOCIAL SCIENCES

Psychology I & II
Sociology

WRITING

College & University Writing

About Research & Education Association

Founded in 1959, Research & Education Association is dedicated to publishing the finest and most effective educational materials— including study guides and test preps—for students in middle school, high school, college, graduate school, and beyond.

Today, REA's wide-ranging catalog is a leading resource for teachers, students, and professionals. Visit *www.rea.com* to see a complete listing of all our titles.

Acknowledgments

We would like to thank Pam Weston, Publisher, for setting the quality standards for production integrity and managing the publication to completion; Larry B. Kling, Vice President, Editorial, for his supervision of revisions and overall direction; Kelli Wilkins, Copywriter, for coordinating development of this edition; Mel Friedman, for his editorial review and revisions; Claudia Petrilli, Graphic Designer, for typesetting this edition; and Christine Saul, Senior Graphic Designer, for designing our cover.

Contents

CHAPTER 1

Integers and

Real Numbers

1.1 Real Numbers and the Number Line

Most of the numbers used in algebra belong to a set called the **real numbers** or **reals**. This set can be represented graphically by the real number line.

Given the number line below, we arbitrarily fix a point and label it with the number 0. In a similar manner, we can label any point on the line with one of the real numbers, depending on its position relative to 0. Numbers to the right of zero are positive, while those to the left are negative. Value increases from left to right, so that if a is to the right of b, it is said to be greater than b.

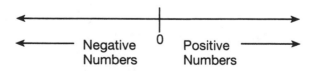

If we now divide the number line into equal segments, we can label the points on this line with real numbers. For example, the point 2 lengths to the left of zero is -2, while the point 3 lengths to the right of zero is + 3 (the + sign is usually assumed, so + 3 is written simply as 3). The number line now looks like this:

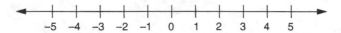

These boundary points represent the subset of the reals known as the **integers**. The set of integers is made up of both the positive and negative whole numbers: $\{\ldots -4, -3, -2, -1, 0, 1, 2, 3, 4, \ldots\}$. Some subsets of integers are:

Natural Numbers or Positive Numbers—the set of integers starting with 1 and increasing: $N = \{1, 2, 3, 4, \ldots\}$.

Whole Numbers—the set of integers starting with 0 and increasing: $W = \{0, 1, 2, 3, \ldots\}$.

Negative Numbers—the set of integers starting with −1 and decreasing: $Z = \{-1, -2, -3, \ldots\}$.

Prime Numbers—the set of positive integers greater than 1 that are divisible only by 1 and themselves: $\{2, 3, 5, 7, 11, \ldots\}$.

Even Integers—the set of integers divisible by 2: $\{\ldots, -4, -2, 0, 2, 4, 6, \ldots\}$.

Odd Integers—the set of integers not divisible by 2: $\{\ldots, -3, -1, 1, 3, 5, 7, \ldots\}$.

Problem Solving Examples:

Graph the following numbers on a number line: −2, 0, 3.

Step 1 is to draw a number line and label it using numbers near the points you are graphing. All points to the left of zero are negative and are always represented by the number with a negative sign (−). All points to the right of zero are positive and may or may not be represented by a positive sign (+).

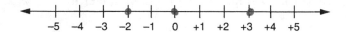

Step 2 is to graph the points on the number line. Points on a number line are graphed using closed circles.

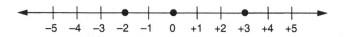

 On the number line below, find and graph the point that is 4 whole numbers to the right of –3.

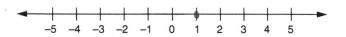

 Step 1 is to find and graph –3.

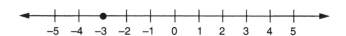

Step 2 is to move 4 whole numbers to the right and graph that point. Therefore, the solution is:

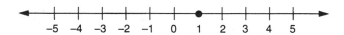

 Using the number line below, graph the solution to 2 + 3.

Step 1 is to graph point 2 on the number line.

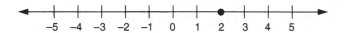

Step 2 is to move 3 units to the right of 2. Since 5 is 3 units to the right of 2, graph 5 on the number line.

Q Using the number line below, graph the solution to −4 + 6.

A Step 1 is to graph point −4 on the number line.

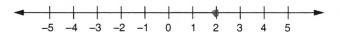

Step 2 is to move 6 units to the right of −4. Since 2 is 6 units to the right of −4, graph 2 on the number line.

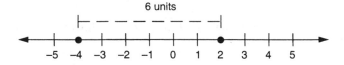

Using the number line below, graph the solution to −5 − (−3).

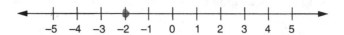

 Step 1 is to graph point −5 on the number line.

Step 2 is to move 3 units to the *right* of −5. In this problem we move to the right of −5 because a negative number is being subtracted. Since −2 is 3 units to the right of −5, graph −2 on the number line.

 Find the sum in the following problem: −8 + (−2).

 Step 1 is to rewrite the problem in the format below.

$$
\begin{array}{r}
-8 \\
+\underline{-2} \\
-10
\end{array}
$$

Step 2 is to add the addends to obtain the sum. The sum is −10.

Find the difference in the following problem:
  $-11 + (+13)$.

Step 1 is to change the sign of the subtrahend.

  $-11 - (-13)$ becomes

  $-11 - (+13)$

Step 2 is to change the problem to an addition problem.

  $-11 - (+13)$ becomes

  $-11 + 13$

Step 3 is to rewrite the problem in the following format.

  -11

  $+\underline{13}$

Step 4 is to add the addends to obtain the sum. The sum is +2. This is also the difference, since we changed the sign of the subtrahend.

Find the product to the following problem:8(0).

The correct answer is 0. When multiplying one or more numbers by zero, the product is always zero.

Find the product in each of the following problems:

  a) $(-6)(4)(1)$ d) $(-1)(-2)(-4)(-1)$

  b) $(-2)(-2)(-2)$ e) $(-2)(-3)(-2)(5)$

  c) $(6)(3)(10)(0)$

Step 1 in each of the problems is to simplify and rewrite as a product of two numbers.

a) $(-6)(4)(1)$ becomes $(-6)(4) \times (1)$ or $(-24)(1)$

b) $(-2)(-2)(-2)$ becomes $(-2)(-2) \times (-2)$ or $(4)(-2)$

c) $(6)(3)(10)(0)$ becomes $(6)(3) \times (10)(0)$ or $(18)(0)$

d) $(-1)(-2)(-4)(-1)$ becomes $(-1)(-2) \times (-4)(-1)$ or $(2)(4)$

e) $(-2)(-3)(-2)(5)$ becomes $(-2)(-3) \times (-2)(5)$ or $(6)(-10)$

Step 2 is to perform the multiplication in each problem.

a) $(-24)(1) = -24$

b) $(4)(-2) = -8$

c) $(18)(0) = 0$

d) $(2)(4) = 8$

e) $(6)(-10) = -60$

Find the product in each of the following problems:

a) $(-6)(1)(3)(2)$ d) $-5 \times -2 \times 11$

b) $(-20)(-3)$ e) 7×11

c) 15×15

In problem a, Step 1 is to simplify and write the problem as a product of two numbers.

$(-6)(1)(3)(2)$ becomes $(-6)(1) \times (3)(2)$ or $(-6)(6)$

Step 2 is to perform the multiplication.

$(-6)(6) = -36$

The product is −36.

In problems b and c, the problems are already expressed as a product of two numbers. The only step is to perform the multiplication.

b) $(-20)(-3) = 60$

The product is 60.

c) $15 \times 15 = 225$

The product is 225.

In problem d, Step 1 is to simplify and write the problem as a product of two numbers.

$-5 \times -2 \times 11$ becomes $(-5)(-2) \times (11)$ or $(10)(11)$

Step 2 is to perform the multiplication.

$(10)(11) = 110$

The product is 110.

In problem e, the problem is already expressed as a product of two numbers. The only step is to perform the multiplication.

$7 \times 11 = 77$

The product is 77.

 Find the quotient in each of the following problems:

a) $-18/3$

b) $99 \div 9$

c) $\dfrac{124}{2}$

d) $7\overline{)35}$

Division can be represented several different ways. Each of the above problems illustrates how division can be represented. The correct answers are problem a) –6, problem b) 11, problem c) 62, and problem d) 5.

 Find the quotient in the following problems:

a) [(8)(–2)]/(–4)   d) [(3)(–7)]/3

b) (18/2)/(15/5)   e) 0/3

c) 141/0

 In problem a, Step 1 is to perform the operation inside the brackets.

(8)(–2) = –16

Step 2 is to rewrite the problem.

(–16)/(–4)

Step 3 is to divide –16 by –4. Since the numerator and denominator are both negative, the answer will be positive.

The quotient is 4.

In problem b, Step 1 is to perform the operation inside the parentheses.

(18/2) = 9 and (15/5) = 3

Step 2 is to rewrite the problem.

9/3

Step 3 is to divide 9 by 3.

The quotient is 3.

In problem c, the quotient is undefined. Any number divided by 0 is always undefined.

In problem d, Step 1 is to perform the operation inside the brackets.

$$(3)(-7) = -21$$

Step 2 is to rewrite the problem.

$$-21/3.$$

Since the numerator is negative and the denominator is positive, the answer will be negative.

The quotient is -7.

In problem e, the quotient is 0. Zero divided by any non-zero number will always be zero.

CHAPTER 2

Fractions

The fraction, a/b, where the **numerator** is a and the **denominator** is b, implies that a is being divided by b. The denominator of a fraction can never be zero since a number divided by zero is not defined. If the numerator is greater than or equal to the denominator, the fraction is called an **improper fraction**. A **mixed number** is the sum of a whole number and a fraction, i.e., $4^3/_8 = 4 + {}^3/_8$.

The number line may be used to show the relationship between integers and fractions. For example, if the interval between 0 and 1 is marked off to form three equal spaces (thirds), then each space so formed is one-third of the total interval. If we move along the number line from 0 toward 1, we will have covered two of the three "thirds" when we reach the second mark. Thus the position of the second mark represents the number ${}^2/_3$.

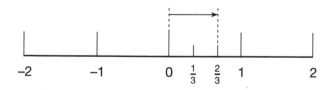

Integers and fractions on the number line

2.1 Simplifying Fractions

To simplify a fraction is to convert it into a form in which the numerator and denominator have no common factor other than 1, e.g.,

$$\frac{12}{18} = \frac{12 \div 6}{18 \div 6} = \frac{2}{3}$$

Problem Solving Examples:

 Simplify the following fraction: $\frac{5}{10}$.

 Step 1 is to find the greatest common divisor (GCD) of 5 and 10. To do this, write out each number as a product of its primes.

$$5 = 5 \times 1 \quad 10 = 5 \times 2$$

The GCD of 5 and 10 is 5.

Step 2 is to divide the numerator and denominator by the GCD.

$$\frac{5}{5} = 1 \quad \frac{10}{5} = 2$$

The correct answer is $\frac{1}{2}$.

 Simplify the following fraction: $\frac{244}{12}$.

 Step 1 is to find the greatest common divisor (GCD) of 244 and 12. To do this, write out each number as products of its primes.

$$244 = 2 \times 2 \times 61 \qquad 12 = 4 \times 3$$

$$= 2 \times 2 \times 3$$

The GCD of 244 and 12 is 4.

Step 2 is to divide the numerator and denominator by the GCD.

$$\frac{244}{4} = 61 \qquad \frac{12}{4} = 3$$

The correct answer is $\frac{61}{3}$ or $20\frac{1}{3}$.

 Simplify the following fraction: $\frac{0}{12}$.

In this problem, the fraction cannot be simplified. Any fraction with 0 in the numerator cannot be simplified. If the numerator and denominator are written as products of their primes, there is no greatest common divisor (GCD).

$$0 = 0 \times 1 \qquad 12 = 4 \times 2$$

$$= 2 \times 2 \times 2$$

Since there is no GCD, the answer is $\frac{0}{12}$, which can be simplified to 0.

2.2 Equivalent Fractions

 Find the lowest common multiple for 7 and 12.

 Step 1 is to list multiples of 7 and 12.

7 = 7, 14, 21, 28, 35, 42, 49, 56, 63, 70, 77, 84

12 = 12, 24, 36, 48, 60, 72, 84, 96, 108, 120, 132, 144

The lowest common multiple of 7 and 12 is 84.

 Find the least common denominator (LCD) for $\frac{6}{8}$ and $\frac{2}{3}$.

 Step 1 is to list multiples of each denominator.

8 = 8, 16, 24, 32, 40

3 = 3, 6, 9, 12, 15, 18, 21, 24, 27, 30

Step 2 is to find the lowest common multiple for both numbers. The lowest common multiple for 8 and 3 is 24. Since 24 is the lowest common multiple, it is also the LCD.

The correct answer is 24.

 Find the lowest common denominator (LCD) for $\frac{5}{6}$, $\frac{4}{21}$, and $\frac{1}{7}$.

 Step 1 is to list multiples of each denominator.

6 = 6, 12, 18, 24, 30, 36, 42, 48, 54, 60, 66, 72

21 = 21, 42, 63, 84, 105

7 = 7, 14, 21, 28, 35, 42, 49, 56, 63, 70

The lowest common multiple of 6, 21, and 7 is 42. Since 42 is the lowest common multiple, it is also the LCD.

Find the equivalent fractions in the following problems:

a) $\dfrac{2}{3} = \dfrac{?}{6} = \dfrac{?}{9}$

b) $\dfrac{4}{5} = \dfrac{?}{20} = \dfrac{?}{10}$

c) $\dfrac{1}{3} = \dfrac{?}{12} = \dfrac{?}{15}$

A Step 1 in each problem is to divide the denominator of the incomplete fraction by the denominator of the complete fractions. The results will become the missing numerators.

Problem a) $\dfrac{2}{3}$ is the complete fraction. $\dfrac{?}{6}$ and $\dfrac{?}{9}$ are the incomplete fractions. Therefore, $6 \div 3 = 2$ and $9 \div 3 = 3$.

Problem b) $\dfrac{4}{5}$ is the complete fraction. $\dfrac{?}{20}$ and $\dfrac{?}{10}$ are the incomplete fractions. Therefore, $20 \div 5 = 4$ and $10 \div 5 = 2$.

Problem c) $\dfrac{1}{3}$ is the complete fraction. $\dfrac{?}{12}$ and $\dfrac{?}{15}$ are the incomplete fractions. Therefore, $12 \div 3 = 4$ and $15 \div 3 = 5$.

Step 2 is to multiply the above results by the numerator in the complete fraction. The results will be the value of the missing numerators.

In problem a) the numerator of the complete fraction is 2. Therefore,

$2 \times 2 = 4$ and $3 \times 2 = 6$.

In problem b) the numerator of the complete fraction is 4. Therefore,

$4 \times 4 = 16$ and $2 \times 4 = 8$.

In problem c) the numerator of the complete fraction is 1. Therefore,

$4 \times 1 = 4$ and $5 \times 1 = 5$.

Step 3 is to rewrite the incomplete fractions using the results above to get the equivalent fraction.

In problem a) the results were 4 and 6. Therefore, $\dfrac{?}{6} = \dfrac{4}{6}$ and $\dfrac{?}{9} = \dfrac{6}{9}$.

Problem b) results were 16 and 8. Therefore, $\dfrac{?}{20} = \dfrac{16}{20}$ and $\dfrac{?}{10} = \dfrac{8}{10}$.

Problem c) results were 4 and 5. Therefore, $\dfrac{?}{12} = \dfrac{4}{12}$ and $\dfrac{?}{15} = \dfrac{5}{15}$.

The correct equivalent fractions are problem a) $\dfrac{4}{6}$ and $\dfrac{6}{9}$, problem b) $\dfrac{16}{20}$ and $\dfrac{8}{10}$, and problem c) $\dfrac{4}{12}$ and $\dfrac{5}{15}$.

2.3   Adding and Subtracting Fractions

To find the sum of two fractions having a common denominator, simply add together the numerators of the given fractions and put this sum over the common denominator.

$$\frac{11}{3} + \frac{5}{3} = \frac{11+5}{3} = \frac{16}{3}$$

Similarly for subtraction,

$$\frac{11}{3} - \frac{5}{3} = \frac{11-5}{3} = \frac{6}{3} = 2$$

To find the sum of the two fractions having different denominators, it is necessary to find the **lowest common denominator (LCD)** of the different denominators using a process called **factoring**.

To **factor** a number means to find two numbers that when multiplied together have a product equal to the original number. These two numbers are then said to be **factors** of the original number. E.g., the factors of 6 are

(1) 1 and 6 since $1 \times 6 = 6$.

(2) 2 and 3 since $2 \times 3 = 6$.

Every number is the product of itself and 1. A **prime factor** is a number that does not have any factors besides itself and 1. This is important when finding the LCD of two fractions having different denominators.

To find the LCD of $^{11}/_6$ and $^5/_{16}$, we must first find the prime factors of each of the two denominators.

$$6 = 2 \times 3$$

$$16 = 2 \times 2 \times 2 \times 2$$

$$\text{LCD} = 2 \times 2 \times 2 \times 2 \times 3 = 48$$

Note that we do not need to repeat the 2 that appears in both the factors of 6 and 16.

Once we have determined the LCD of the denominators, each of the fractions must be converted into equivalent fractions having the LCD as a denominator.

Rewrite 11/6 and 5/16 to have 48 as their denominators.

$$6 \times ? = 48 \quad 16 \times ? = 48$$

$$6 \times 8 = 48 \quad 16 \times 3 = 48$$

If the numerator and denominator of each fraction is multiplied (or divided) by the same number, the value of the fraction will not change. This is because a fraction b/b, b being any number except zero, is equal to the multiplicative identity, 1.

Therefore,

$$\frac{11}{6} \times \frac{8}{8} = \frac{88}{48} \qquad \frac{5}{16} \times \frac{3}{3} = \frac{15}{48}$$

We may now find

$$\frac{11}{6} + \frac{5}{16} = \frac{88}{48} + \frac{15}{48} = \frac{103}{48}$$

Similarly for subtraction,

$$\frac{11}{6} - \frac{5}{16} = \frac{88}{48} - \frac{15}{48} = \frac{73}{48}$$

Problem Solving Examples:

 Find the solution to the following problem:

$$-\frac{3}{7}-\frac{2}{7}.$$

 Step 1 is to subtract the numerators.

$$-3-2=-5$$

Since the denominators in $-\frac{3}{7}$ and $-\frac{2}{7}$ are equal, keep the common denominator.

The correct answer is $-\frac{5}{7}$.

 Find the solution to the following problem:

$$\frac{1}{3}+\frac{1}{2}.$$

 The denominators in $\frac{1}{3}$ and $\frac{1}{2}$ are not equal. In this type of problem use the least common denominator (LCD) to add the fractions.

Step 1 is to list multiples of each denominator.

3 = 3, 6, 9, 12, 15

2 = 2, 4, 6, 8, 10

The LCD is 6.

Step 2 is to divide the LCD by each denominator.

The denominator for $\frac{1}{3}$ is 3 so, 6 ÷ 3 = 2.

The denominator for $\frac{1}{2}$ is 2 so, 6 ÷ 2 = 3.

Step 3 is to multiply those results by each numerator to get the new numerator.

The result was 2 and the numerator for $\frac{1}{3}$ is 1. Therefore, 2 × 1 = 2. The new numerator is 2.

The result was 3 and the numerator for $\frac{1}{2}$ is 1. Therefore, 3 × 1 = 3. The new numerator is 3.

Step 4 is to write $\frac{1}{3}$ and $\frac{1}{2}$ in terms of their LCD by rewriting each fraction using the new numerator.

$$\frac{1}{3} = \frac{2}{6}$$

$$\frac{1}{2} = \frac{3}{6}$$

Step 5 is to add the numerators.

2 + 3 = 5

The correct answer is $\frac{5}{6}$.

Find the solution to the following problem:

$$\frac{1}{3} + \frac{3}{6} - \frac{2}{9}.$$

A The denominators in $\frac{1}{3}$, $\frac{3}{6}$, and $\frac{2}{9}$ are not equal, so you must determine the least common denominator (LCD) to perform the operations.

Step 1 is to list multiples of each denominator.

$3 = 3, 6, 9, 12, 15, 18, 21$

$6 = 6, 12, 18, 24, 30$

$9 = 9, 18, 27, 36$

The LCD is 18.

Step 2 is to divide the LCD by each denominator.

The denominator for $\frac{1}{3}$ is 3. Therefore, $18 \div 3 = 6$.

The denominator for $\frac{3}{6}$ is 6. Therefore, $18 \div 6 = 3$.

The denominator for $\frac{2}{9}$ is 9. Therefore, $18 \div 9 = 2$.

Step 3 is to multiply those results by each numerator to get the new numerator.

The result was 6 and the numerator for $\frac{1}{3}$ is 1. Therefore, $1 \times 6 = 6$. The new numerator is 6.

The result was 3 and the numerator for $\frac{3}{6}$ is 3. Therefore, 3×3 = 9. The new numerator is 9.

The result was 2 and the numerator for $\frac{2}{9}$ is 2. Therefore, 2×2 = 4. The new numerator is 4.

Step 4 is to write $\frac{1}{3}$, $\frac{3}{6}$, and $\frac{2}{9}$ in terms of their LCD by rewriting each fraction using the new numerator.

$$\frac{1}{3} = \frac{6}{18}$$

$$\frac{3}{6} = \frac{9}{18}$$

$$\frac{2}{9} = \frac{4}{18}$$

Step 5 is to perform the operations of the numerators.

$$6 + 9 - 4 = 11$$

The correct answer is $\frac{11}{18}$

 Find the solution to the following problem:

$$\left[-\frac{31}{48} + \left(-\frac{11}{48}\right)\right] - \frac{6}{8}.$$

A Step 1 is to perform the operation inside the brackets. Since the denominators in $-\dfrac{31}{48}$ and $-\dfrac{11}{48}$ are equal, keep the common denominator. Next, add the numerators.

$$-\frac{31}{48} + \left(-\frac{11}{48}\right) = -\frac{42}{48}$$

Rewritten, the problem now reads

$$-\frac{42}{48} - \frac{6}{8}.$$

Step 2 is to find the least common denominator (LCD) of $-\dfrac{42}{48}$ and $-\dfrac{6}{8}$. To do this, list multiples of each denominator.

$$48 = 48, 96, 144$$

$$8 = 8, 16, 24, 32, 40, 48, 56, 64$$

The LCD is 48.

Step 3 is to divide the LCD by each denominator.

The denominator for $-\dfrac{42}{48}$ is 48. Therefore, $48 \div 48 = 1$.

The denominator for $-\dfrac{6}{8}$ is 8. Therefore, $48 \div 8 = 6$.

Step 4 is to multiply those results by each numerator to get the new numerator.

The result was 1 and the numerator for $-\dfrac{42}{48}$ is 42. Therefore, $42 \times 1 = 42$. The new numerator is 42.

The result was 6 and the numerator for $-\dfrac{6}{8}$ is 6. Therefore, $6 \times 6 = 36$. The new numerator is 36.

Step 5 is to rewrite $-\dfrac{42}{48}$ and $-\dfrac{6}{8}$ in terms of their LCD by rewriting each fraction using the new numerator.

$$-\frac{42}{48} = -\frac{42}{48}$$

$$-\frac{6}{8} = \frac{36}{48}$$

Next, combine the numerators.

$$-42 - 36 = -78$$

The correct answer is $-\dfrac{78}{48}$, or simplified, $-\dfrac{13}{8}$.

2.4   Multiplying and Dividing Fractions

To find the product of two or more fractions, simply multiply the numerators of the given fractions to find the numerator of the product and multiply the denominators of the given fractions to find the denominator of the product. E.g.,

$$\frac{2}{3} \times \frac{1}{5} \times \frac{4}{7} = \frac{2 \times 1 \times 4}{3 \times 5 \times 7} = \frac{8}{105}$$

To find the quotient of two fractions, simply invert the divisor and multiply. E.g.,

$$\frac{8}{9} \div \frac{1}{3} = \frac{8}{9} \times \frac{3}{1} = \frac{24}{9} = \frac{8}{3}$$

Problem Solving Examples:

 Which of the following rules about multiplying fractions is true?

a) $\dfrac{a}{c} \times \dfrac{b}{d} = \dfrac{a(b)}{b+c}$

b) $\dfrac{a}{c} \times \dfrac{b}{d} = \dfrac{a(b)}{c(d)}$

c) $\dfrac{a}{c} \times \dfrac{b}{d} = \dfrac{a+b}{c+d}$

 The correct answer is rule b. The rule for multiplying fractions is to multiply the numerators, then multiply the denominators. Choices rule a and rule c are incorrect and do not represent any mathematical rule pertaining to fractions.

 Find the solution to the following problem:

$$\frac{5}{6} \times \frac{2}{3}.$$

 Step 1 is to multiply the numerators.

$5 \times 2 = 10$

Step 2 is to multiply the denominators.

$6 \times 3 = 18$

The correct answer is $\dfrac{10}{18}$, simplified as $\dfrac{5}{9}$.

Which of the following rules about dividing fractions is true?

a) $\dfrac{a}{c} \div \dfrac{b}{d} = \dfrac{a(d)}{b(c)}$

b) $\dfrac{a}{c} \div \dfrac{b}{d} = \dfrac{a-b}{c-d}$

c) $\dfrac{a}{c} \div \dfrac{b}{d} = \dfrac{a+b}{c+d}$

The correct answer is rule a. The rule for dividing fractions is to take the reciprocal of the divisor, then multiply the numerators, then multiply the denominators. Rules b and c are incorrect and do not represent any mathematical rule pertaining to fractions.

Find the solution to the following problem:

$\dfrac{9}{15} \div \dfrac{0}{3}$.

Step 1 is to take the reciprocal of (or invert) the divisor.

$\dfrac{9}{15} \div \dfrac{3}{0}$

Step 2 is to change the operation to multiplication.

$\dfrac{9}{15} \times \dfrac{3}{0}$

Step 3 is to multiply the numerators and denominators.

$9 \times 3 = 27$

$15 \times 0 = 0$

Step 4 is to rewrite the fraction.

$\dfrac{27}{0}$, which is undefined.

Any fraction whose denominator is 0 is undefined.

Find the products in the following problems. Simplify the answers.

a) $\dfrac{1}{2} \times \dfrac{7}{16}$

b) $\dfrac{5}{8} \times \dfrac{11}{12}$

c) $\dfrac{12}{14} \times \dfrac{12}{14}$

d) $\dfrac{53}{61} \times \dfrac{1}{2}$

Step 1 in each problem is to multiply the numerators.

a) $1 \times 7 = 7$

b) $5 \times 11 = 55$

c) $12 \times 12 = 144$

d) $53 \times 1 = 53$

Step 2 is to multiply the denominators.

a) $2 \times 16 = 32$

b) $8 \times 12 = 96$

c) $14 \times 14 = 196$

d) $61 \times 2 = 122$

Step 3 is to rewrite the fractions to get the product. Simplify if necessary.

a) $\dfrac{7}{32}$

b) $\dfrac{55}{96}$

c) $\dfrac{144}{196}$ or $\dfrac{36}{49}$

d) $\dfrac{53}{122}$

 Find the quotients in the following problems. Simplify the answers.

a) $\dfrac{8}{16} \div \dfrac{4}{1}$

b) $\dfrac{5}{8} \div \dfrac{3}{0}$

c) $\dfrac{5}{5} \div \dfrac{2}{1}$

d) $\dfrac{9}{11} \div \dfrac{10}{4}$

A All but problem b can be solved. For b, no operation can be performed, as division by zero is undefined. $\dfrac{5}{8}$ divided by an undefined quantity yields no solution.

Step 1 for a, c, and d is to take the reciprocal of (or invert) the divisor.

a) $\dfrac{8}{16} \div \dfrac{1}{4}$

c) $\dfrac{5}{5} \div \dfrac{1}{2}$

d) $\dfrac{9}{11} \div \dfrac{4}{10}$

Step 2 is to change the operation to multiplication.

a) $\dfrac{8}{16} \times \dfrac{1}{4}$

c) $\dfrac{5}{5} \times \dfrac{1}{2}$

d) $\dfrac{9}{11} \times \dfrac{4}{10}$

Step 3 is to multiply the numerators.

a) $8 \times 1 = 8$

c) $5 \times 1 = 5$

d) $9 \times 4 = 36$

Step 4 is to multiply the denominators.

a) $16 \times 4 = 64$

c) $5 \times 2 = 10$

d) $11 \times 10 = 110$

Step 5 is to rewrite the fractions to get the quotient. Simplify if necessary.

a) $\dfrac{8}{64}$ or $\dfrac{1}{8}$

c) $\dfrac{5}{10}$ or $\dfrac{1}{2}$

d) $\dfrac{36}{110}$ or $\dfrac{18}{55}$

 Find the solution to the following problem:

$$\frac{4+5}{3+7} \times \frac{3}{7}.$$

 Step 1 is to perform the operations in the first fraction. Note that $\dfrac{4+5}{3+7}$ is different from $\dfrac{4}{3} + \dfrac{5}{7}$.

$4 + 5 = 9$

$3 + 7 = 10$

Step 2 is to rewrite the problem using the new fraction.

$$\frac{9}{10} \times \frac{3}{7}$$

Step 3 is to multiply the numerators.

$9 \times 3 = 27$

Step 4 is to multiply the denominators.

$10 \times 7 = 70$

The correct answer is $\dfrac{27}{70}$.

 Find the solution to the following problem. Simplify the answer.

$$\frac{1}{4} \times \frac{6}{8} \div \frac{2}{3}$$

 In mathematics, multiplication precedes division in the order of operations. Therefore, the first step is to multiply $\dfrac{1}{4}$ and $\dfrac{6}{8}$.

$$\frac{1}{4} \times \frac{6}{8}$$

The answer is $\dfrac{6}{32}$.

Next, perform the division.

$$\frac{6}{32} \div \frac{2}{3}$$

Invert the divisor.

$$\frac{6}{32} \div \frac{3}{2}$$

Change the operation to multiplication.

$$\frac{6}{32} \times \frac{3}{2}$$

Next, multiply the numerators.

$$6 \times 3 = 18$$

Next, multiply the denominators.

$$32 \times 2 = 64$$

The correct answer is $\frac{18}{64}$, or $\frac{9}{32}$.

2.5   Mixed Numbers

To change a mixed number to an improper fraction, simply multiply the whole number by the denominator of the fraction and add the numerator. This product becomes the numerator of the result and the denominator remains the same. E.g.,

$$5\frac{2}{3} = \frac{(5 \times 3) + 2}{3} = \frac{15 + 2}{3} = \frac{17}{3}$$

To change an improper fraction to a mixed number, simply divide the numerator by the denominator. The remainder becomes the numerator of the fractional part of the mixed number, and the denominator remains the same. E.g.,

$$\frac{35}{4} = 35 \div 4 = 8\frac{3}{4}$$

To check your work, change your result back to an improper fraction to see if it matches the original fraction.

Problem Solving Examples:

 Write $\dfrac{9}{6}$ as a mixed number.

 Step 1 is to find the greatest common divisor (GCD) for $\dfrac{9}{6}$. To do this, write 9 and 6 as products of their primes.

$$9 = 3 \times 3 \quad 6 = 3 \times 2$$

The GCD of 9 and 6 is 3.

Step 2 is to divide the numerator and denominator by the GCD.

$$\frac{9}{3} = 3 \quad \frac{6}{3} = 2$$

$\dfrac{9}{6}$ is simplified as $\dfrac{3}{2}$.

Step 3 is to write $\dfrac{3}{2}$ as a mixed number by dividing 2 into 3.

$$\frac{3}{2} = 1\frac{1}{2}$$

Thus, $\dfrac{9}{6}$, simplified and written as a mixed number, is $1\dfrac{1}{2}$.

Simplify the following fraction: $2\frac{12}{8}$.

 Step 1 is to convert $2\frac{12}{8}$ into an improper fraction. The whole number "2" gets multiplied by the denominator "8."

$2 \times 8 = 16$

Next, add the numerator to the previous result.

$16 + 12 = 28$

The improper fraction is $\frac{28}{8}$.

Step 2 is to simplify $\frac{28}{8}$.

Find the greatest common divisor (GCD) of 28 and 8. This is done by writing the numbers as products of their primes.

$28 = 14 \times 2 \qquad 8 = 4 \times 2$

$28 = 7 \times 2 \times 2 \qquad 8 = 2 \times 2 \times 2$

The GCD of 28 and 8 is 4.

Next, divide the numerator and denominator by the GCD.

$\frac{28}{4} = 7 \qquad\qquad \frac{8}{4} = 2$

Therefore, $2\frac{12}{8}$ simplified is $\frac{7}{2}$. $\frac{7}{2}$ can also be written as a mixed number, $3\frac{1}{2}$.

Find the solution to the following problem:

$$5\frac{1}{8} \times 2\frac{2}{3}.$$

 Step 1 is to convert $5\frac{1}{8}$ to an improper fraction. The whole number "5" gets multiplied by the denominator "8."

$5 \times 8 = 40$

Next, add the numerator to the previous result.

$1 + 40 = 41$

The improper fraction of $5\frac{1}{8}$ is $\frac{41}{8}$.

Step 2 is to convert $2\frac{2}{3}$ to an improper fraction.

The whole number "2" gets multiplied by the denominator "3."

$2 \times 3 = 6$

Next, add the numerator to the previous result.

$2 + 6 = 8$

The improper fraction of $2\frac{2}{3}$ is $\frac{8}{3}$.

Step 3 is to multiply $\frac{41}{8}$ and $\frac{8}{3}$.

$$\frac{41}{8} \times \frac{8}{3}$$

To do this, multiply the numerators and denominators.

$$\frac{41}{8} \times \frac{8}{3} = \frac{328}{24}$$

The correct answer is $\frac{328}{24}$. To write $\frac{328}{24}$ as a mixed number,

divide 24 into 328. The mixed number is $13\frac{16}{24}$, which equals $13\frac{2}{3}$

when simplified.

The solution is $13\frac{2}{3}$.

 Find the solution to the following problem:

$$\frac{1}{2} - 7\frac{1}{2}.$$

 Since $\frac{1}{2}$ is not a mixed number, Step 1 is to convert $7\frac{1}{2}$ to an improper fraction.

The whole number "7" gets multiplied by the denominator "2."

$7 \times 2 = 14$

Next, add the numerator to the previous result.

$1 + 14 = 15$

The improper fraction of $7\frac{1}{2}$ is $\frac{15}{2}$.

Step 2 is to rewrite the problem.

$$\frac{1}{2} - \frac{15}{2}$$

Step 3 is to perform the subtraction.

$$\frac{1}{2} - \frac{15}{2} = -\frac{14}{2}$$

The correct answer is $-\frac{14}{2}$. To write $-\frac{14}{2}$ as a mixed number,

divide 2 into 14. The mixed number is $-\frac{7}{1}$, or -7 when simplified.

CHAPTER 3

Decimals

When we divide the denominator of a fraction into its numerator, the result is a **decimal**. The decimal is based upon a fraction with a denominator of 10, 100, 1,000, ... and is written with a **decimal point**. Whole numbers are placed to the left of the decimal point where the first place to the left is the units place; the second to the left is the tens; the third to the left is the hundreds, etc. The fractions are placed on the right where the first place to the right is the tenths; the second to the right is the hundredths, etc.

In the decimal system, each digit position in a number has ten times the value of the position adjacent to it on the right. For example, in the number 11, the 1 on the left is said to be in the "tens place," and its value is 10 times as great as that of the 1 on the right. The 1 on the right is said to be in the "units place," with the understanding that the term "unit" in our system refers to the numeral 1. Thus the number 11 is actually a coded symbol which means "one ten plus one unit." Since ten plus one is eleven, the symbol 11 represents the number 11.

3.1   Changing Fractions to Decimals

 Using the number line below, match the letters with the following decimals: 1.00, –0.25, –0.75, 0.50, –1.25.

 The correct answers are:

   A   –1.25

   B   –0.75

   C   –0.25

   D   0.50

   E   1.00

All fractions have an associated decimal value. To find the decimal value of a fraction, treat the fraction like a division problem.

$$A = -\frac{5}{4} \text{ and } -5 \div 4 = -1.25$$

$$B = -\frac{3}{4} \text{ and } -3 \div 4 = -0.75$$

$$C = -\frac{1}{4} \text{ and } -1 \div 4 = -0.25$$

$$D = \frac{2}{4} \text{ and } 2 \div 4 = 0.50$$

$$E = \frac{4}{4} \text{ and } 4 \div 4 = 1.00$$

Using the number line below as a reference, which statements are true?

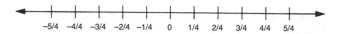

a) $-\dfrac{1}{4} > -0.50$

b) $\dfrac{5}{4} = 1.25$

c) $-\dfrac{1}{2} < 0.25$

d) $-\dfrac{3}{4} = -3.75$

 The true statements are statement a, statement b, and statement c.

Statement a is true because $-\dfrac{1}{4}$ is the same as $-1 \div 4 = -0.25$. Since -0.25 is greater than -0.50, the statement is true.

Statement b is true because $\dfrac{5}{4}$ is the same as $5 \div 4 = 1\dfrac{1}{4}$ or 1.25.

Statement c is true because $-\dfrac{1}{2}$ is the same as $-1 \div 2 = -0.50$. Since -0.50 is less than 0.25, the statement is true.

Statement d is incorrect because $-\dfrac{3}{4}$ is the same as $-3 \div 4 = -0.75$. Since -0.75 does not equal -3.75, the statement is false.

Convert the following fraction to a decimal: $\dfrac{1}{8}$.

 Step 1 is to rewrite the fraction as a division problem.

$$\dfrac{1}{8} = 1 \div 8$$

Step 2 is to solve the division problem.

$$1 \div 8 = 8\overline{)1.00}$$

$$
\begin{array}{r}
0.125 \\
8\overline{)1.000} \\
\underline{8} \\
20 \\
\underline{16} \\
40 \\
\underline{40} \\
0
\end{array}
$$

The correct answer is 0.125

 Convert the following fraction to a decimal: $-\dfrac{5}{9}$.

 Step 1 is to rewrite the fraction as a division problem.

$$-\dfrac{5}{9} = -5 \div 9$$

Step 2 is to solve the division problem.

$$-5 \div 9 = 9\overline{)5.00}$$

$$\begin{array}{r} 0.555 \\ 9\overline{)5.000} \\ \underline{4.5} \\ 50 \\ \underline{45} \\ 50 \\ \underline{45} \\ 5 \end{array}$$

The correct answer is –0.555. This is known as a repeating decimal because if –5 ÷ 9 is continuously divided, the decimal will repeat to infinity— –0.555555....

Convert the following fraction to a decimal: $3\frac{3}{4}$.

Step 1 is to rewrite the fraction as a division problem, using only the fraction, not the integer.

$$\frac{3}{4} = 3 \div 4$$

Step 2 is to solve the division problem.

$$3 \div 4 = 4\overline{)3.00}$$

$$\begin{array}{r} 0.75 \\ 4\overline{)3.00} \\ \underline{2.8} \\ 20 \\ \underline{20} \\ 0 \end{array}$$

Step 3 is to add the decimal to the integer.

$$0.75 + 3 = 3.75$$

The correct answer is 3.75.

3.2 Changing Decimals to Fractions

Using the number line below, match the letters with the following fractions: $\dfrac{1}{2}, \dfrac{8}{4}, -\dfrac{15}{10}, -2\dfrac{1}{2}$.

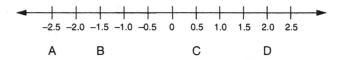

The correct answers are:

A $-2\dfrac{1}{2}$

B $-\dfrac{15}{10}$

C $\dfrac{1}{2}$

D $\dfrac{8}{4}$

Most decimals have an associated fraction. To find the fraction of a decimal, Step 1 is to write the decimal in terms of the fraction. If there is an integer, leave it in place. If the decimal only goes out to a tenth, make the fraction over 10; if the decimal goes to the hundredths place, make the fraction over 100, and so on.

$$A = -2.5 = -2 + \dfrac{5}{10} \qquad C = 0.5 = \dfrac{5}{10}$$

$$B = -1.5 = -1 + \frac{5}{10} \qquad D = 2.0 = 2 + \frac{0}{10}$$

Step 2 is to rewrite the fraction, using the integer. Do not add the integer, simply join the integer and the fraction.

$$A = -2 + \frac{5}{10} = -2\frac{5}{10} \qquad C = \frac{5}{10}$$

$$B = -1 + \frac{5}{10} = -1\frac{5}{10} \qquad D = 2 + \frac{0}{10} = 2\frac{0}{10}$$

Step 3 is to simplify the mixed numbers.

$$A = -2\frac{1}{2} \qquad\qquad C = \frac{1}{2}$$

$$B = -1\frac{1}{2} \qquad\qquad D = 2$$

 Convert the following decimal to a fraction: 4.95.

 Step 1 is to write the decimal in terms of the fraction one-hundredth. Leave the integer "4" in place.

$$4 \text{ and } \frac{95}{100}$$

Step 2 is to simplify the fraction $\frac{95}{100}$.

$$\frac{95}{100} = \frac{19}{20}$$

Step 3 is to rewrite the fraction, using the integer. Simply join the integer and the fraction.

$$4 \text{ and } \frac{19}{20} = 4\frac{19}{20}$$

The correct answer is $4\frac{19}{20}$.

 Convert the following decimal to a fraction: 12.234.

 Step 1 is to write the decimal in terms of the fraction one-thousandth. Leave the integer "12" in place.

$$12 \text{ and } \frac{234}{1000}$$

Step 2 is to simplify the fraction $\frac{234}{1000}$.

$$\frac{234}{1000} = \frac{117}{500}$$

Step 3 is to rewrite the fraction, using the integer. Simply join the integer and the fraction.

$$12 \text{ and } \frac{117}{500} = 12\frac{117}{500}$$

The correct answer is $12\dfrac{117}{500}$.

 Convert the following decimal to a fraction: $6.333\overline{3}$.

Not all decimals can be converted with the methods used above. These decimals are repeating decimals. To convert repeating decimals into fractions, use the types of fractions that form repeating decimals. Some are listed below.

$$\frac{1}{3} = 0.333\overline{3} \qquad \frac{1}{6} = 0.1666\overline{6} \qquad \frac{1}{9} = 0.111\overline{1}$$

$$\frac{2}{3} = 0.666\overline{6} \qquad \frac{5}{6} = 0.8333\overline{3} \qquad \frac{2}{9} = 0.222\overline{2}$$

$$\frac{4}{9} = 0.444\overline{4}$$

Step 1 is to separate the integer and the fraction.

$$6.333\overline{3} = 6 + 0.333\overline{3}$$

Step 2 is to determine which fraction forms the decimal 0.3333.

$$0.333\overline{3} = \frac{1}{3}$$

Step 3 is to rewrite the fraction, using the integer. Simply join the integer and the fraction.

$$6 \text{ and } \frac{1}{3} = 6\frac{1}{3}$$

The correct answer is $6\dfrac{1}{3}$.

Q Convert the following decimal to a fraction: –45.45.

A Step 1 is to write the decimal in terms of the fraction one-hundredth. Leave the integer "45" in place.

$$-45 \text{ and } \frac{45}{100}$$

Step 2 is to simplify the fraction $\frac{45}{100}$.

$$\frac{45}{100} = \frac{9}{20}$$

Step 3 is to rewrite the fraction, using the integer. Simply join the integer and the fraction.

$$-45 \text{ and } \frac{9}{20} = -45\frac{9}{20}$$

The correct answer is $-45\frac{9}{20}$.

3.3   Rounding Decimals

Q Round the following decimal to the nearest hundred: 134.22.

A Step 1 is to determine the digit that will be rounded.

1̌3̌4.22

Since the digit "1" is in the hundreds' place, it will be rounded.

Step 2 is to locate the digit to the right of "1."

134.22
ˇˇ

The digit "3" is to the right of "1." Set all digits to the right of "3" equal to "0."

130.00
ˇˇ

Step 3 is to determine if the decimal will be rounded up or down. Since "3" is less than 5, the decimal will be rounded down. To do this, set "3" equal to "0."

100.00
ˇˇ

The correct answer is 100.

 Round the following decimal to the nearest tenth: 5.719.

 Step 1 is to determine the digit that will be rounded.

5.719
ˇˇ

Since the digit "7" is in the tenths' place, it will be rounded.

Step 2 is to locate the digit to the right of "7."

5.719
ˇˇ

The digit "1" is to the right of "7." Set all digits to the right of "1" equal to "0."

5.710
ˇˇ

Step 3 is to determine if the decimal will be rounded up or down. Since "1" is less than 5, the decimal will be rounded down. To do this, set "1" equal to "0."

5.7 0 0
  ∨ ∨

The correct answer is 5.7.

 Round the following decimal to the nearest thousandth: 0.9196.

 Step 1 is to determine the digit that will be rounded.

0.9196
   ∨ ∨

Since the digit "9" is in the thousandths' place, it will be rounded.

Step 2 is to locate the digit to the right of "9."

0.9196
   ∨ ∨

The digit "6" is to the right of "9." In this case, there are no digits to the right of "6" to set equal to "0."

0.9196
   ∨ ∨

Step 3 is to determine if the decimal will be rounded up or down. Since "6" is greater than or equal to 5, the decimal will be rounded up. To do this, set "6" equal to "0," and increase "9" by one. Since "9" increased by one is ten, carry the addition over to the remaining decimal places.

0.9200
   ∨ ∨

The correct answer is 0.920.

3.4   Adding and Subtracting Decimals

Find the solution to the following operation.

6.11 + 3.251

Step 1 is to align the decimal point of each number.

6.11

+ 3.251

Step 2: since 3.251 has a digit in the thousandths' place, add a "0" in the thousandths' place in 6.11.

6.110

+ 3.251

Step 3 is to perform the addition.

6.110

+ 3.251

9.361

The solution is 9.361.

Find the solution to the following operation.

0.0598 + 0.000031

Step 1 is to align the decimal point of each number.

0.0598

+ 0.000031

Step 2: since 0.000031 has a digit in the hundred thousandths' and millionths' places, add a "0" in those places in 0.0598.

0.059800

+ 0.000031

Step 3 is to perform the addition.

0.059800

+ 0.000031

0.059831

The solution is 0.059831.

 Find the solution to the following operation.

9.002 – 3.549

 Step 1 is to align the decimals of each number.

9.002

– 3.549

Step 2 is to perform the subtraction.

9.002

– 3.549

5.453

The solution is 5.453.

 Find the solution to the following operation.

135.98 – (–0.31)

 Step 1 is to align the decimal point of each number.

135.98

– (–0.31)

Step 2 is to change the operation to an addition problem.

135.98

+   0.31

Step 3 is to perform the addition.

135.98

+   0.31

136.29

The solution is 136.29.

 Find the solution to the following operation. Round the solution to the nearest whole number.

9,242,985.45212 + 432,885.93311

 Step 1 is to align the decimal point of each number.

9,242,985.45212

+   432,885.93311

Step 2 is to perform the addition.

9,242,985.45212

+   432,885.93311

9,675,871.38523

The solution is 9,675,871.38523. Rounded to the nearest whole number, the solution becomes 9,675,871.

 Find the solution to the following operation. Round the solution to the nearest ten thousandth:

0.00031 – 0.0000043

Step 1 is to align the decimal point of each number.

0.00031

– 0.0000043

Step 2: since 0.0000043 has a digit in the millionths' and ten millionths' places, add a "0" in those places in 0.00031.

0.0003100

– 0.0000043

Step 3 is to perform the subtraction.

0.0003100

– 0.0000043

– 0.0003057

The solution is 0.0003057. Rounded to the nearest ten thousandth, the solution becomes 0.0003.

3.5 Multiplying and Dividing Decimals

Find the solution to the following operation.

–7.5 × 2.8

Step 1 is to align the decimals of the numbers together.

–7.5

× 2.8

Step 2 is to perform the operation. Since there is one decimal place in –7.5 and one decimal place in 2.8, the result will have two decimal places.

−7.5

× 2.8

−21.00

The solution is −21.0.

 Find the solution to the following operation.

0.543 × 0.1012

 Step 1 is to align the decimals of the numbers together.

0.543

× 0.1012

Step 2: since 0.1012 has digits in the thousandths' and ten thousandths' places, add "0" to those places in 0.543.

0.5430

× 0.1012

Step 3 is to perform the operation. Since there are four decimal places in 0.5430 and four decimal places in 0.1012, the result will have eight decimal places.

0.5430

× 0.1012

0.05495160

The solution is 0.05495160.

 Find the solution to the following operation.

8.8 ÷ 1.2

Step 1 is to set up the operation in the format below.

1.2)8.8

Step 2 is to convert the divisor into a whole number. To do this, move the decimal point one place to the right.

1.2 = 12

Step 3, move the decimal point of the dividend the same number of places.

8.8 = 88

Step 4, perform the operation.

```
       7.33
  12)88
     84
     ——
     40
     36
     ——
     40
     36
     ——
```

The solution is $7.3\overline{3}$. This is a repeating decimal. A repeating decimal has a repeating digit that will repeat to infinity. The repeating digit is the number(s) with the bar over the top.

Find the solution to the following operation.

59.2 × 62.13

Step 1 is to align the decimals of the numbers together.

59.2   (1 decimal place)

× 62.13   (2 decimal places)

Step 2 is to determine the number of decimal places in the product.

1 decimal place + 2 decimal places = 3 decimal places

Step 3; since 62.13 has a digit in the hundredths' place, add "0" to that place in 59.2.

$$59.20$$

$$\times \underline{62.13}$$

Step 4 is to perform the operation.

$$59.20$$

$$\times \underline{62.13}$$

$$3678.096$$

The solution is 3678.096.

Find the solution to the following operation.

0.004 × 0.006

Step 1 is to align the decimals of the numbers together.

0.004   (3 decimal places)

× <u>0.006</u>   (3 decimal places)

Step 2 is to determine the number of decimal places in the product.

3 decimal places + 3 decimal places = 6 decimal places

Step 3 is to perform the operation.

0.004

× 0.006

0.000024

The solution is 0.000024.

 Find the solution to the following operation.

127.5 ÷ 5.0

 Step 1 is to set up the operation in the format below.

$$5.0\overline{)127.5}$$

Step 2 is to convert the divisor into a whole number. To do this, move the decimal point one place to the right.

5.0 = 50

Step 3 is to move the decimal point of the dividend the same number of places.

127.5 = 1275

Step 4 is to perform the operation.

```
      25.5
5)127.50
  100
  ---
   275
   250
   ---
    250
    250
    ---
      0
```

The solution is 25.5.

Find the solution to the following operation.

0.0006 ÷ 0.0002

Step 1 is to set up the operation in the format that follows:

$$0.0002\overline{)0.0006}$$

Step 2 is to convert the divisor into a whole number. To do this, move the decimal point four places to the right.

0.0002 = 2

Step 3 is to move the decimal point of the dividend the same number of places.

0.0006 = 6

Step 4 is to perform the operation.

$$
\begin{array}{r}
3 \\
2\overline{)6} \\
6 \\
\hline
0
\end{array}
$$

The solution is 3.

3.6   Mixed Problems

Round the following decimal to the nearest hundredth. Write the decimal as a fraction.

–0.7523

Step 1 is to round –0.7523 to the nearest hundredth. To do this, find the digit in the hundredths' place and determine if the decimal should be rounded up or down.

−0.7523 rounds down to −0.7500

Step 2 is to convert −0.75 to a fraction. To do this, write the decimal in terms of the fraction one-hundredths.

$$-0.75 = -\frac{75}{100}$$

Step 3 is to simplify $-\frac{75}{100}$.

$$-\frac{75}{100} = -\frac{3}{4}$$

The correct answer is $-\frac{3}{4}$.

 Find the solution to the following problem.

$$\frac{9}{10} - (1.56 - 0.87)$$

 Step 1 is to convert $\frac{9}{10}$ to a decimal.

$$\frac{9}{10} = 0.90$$

Step 2 is to perform the operations inside the parentheses.

$$\begin{array}{r} 1.56 \\ -\ 0.87 \\ \hline 0.69 \end{array}$$

Step 3 is to subtract 0.69 from 0.90.

0.90

– 0.69

0.21

The correct answer is 0.21.

 Find the solution to the following operation.

$0.33\overline{3} \div 3$

 Step 1: since $0.33\overline{3}$ is a repeating decimal, to get an answer, the decimal must be converted to a fraction.

$$0.33\overline{3} = \frac{1}{3}$$

Step 2 is to rewrite the problem.

$$\frac{1}{3} \div 3$$

Step 3 is to invert the divisor and change the operation to multiplication.

$$\frac{1}{3} \times \frac{1}{3}$$

Step 4 is to perform the operation.

$$\frac{1}{3} \times \frac{1}{3} = \frac{1}{9}$$

The solution is $\frac{1}{9}$ or $0.11\overline{1}$.

Round the following decimals to the nearest tenth and then perform the operation.

1.48 – 3.77

 Step 1 is to round 1.48 and 3.77 to the nearest tenth.

1.48 rounds to 1.5.

3.77 rounds to 3.8.

Step 2 is to rewrite the problem.

$$1.5$$
$$\underline{-\ 3.8}$$

Step 3 is to perform the operation.

$$1.5$$
$$\underline{-\ 3.8}$$
$$-\ 2.3$$

The correct answer is –2.3.

CHAPTER 4

Ratio and Proportions

The ratio of two numbers x and y written $x : y$ is the fraction x/y where $y \neq 0$. A ratio compares x to y by dividing one by the other. Therefore, in order to compare ratios, simply compare the fractions.

The results of observation or measurement often must be compared with some standard value in order to have any meaning. For example, to say that a man can read 400 words per minute has little meaning as it stands. However, when his rate is compared to the 250 words per minute of the average reader, one can see that he reads considerably faster than the average reader. How much faster? To find out, his rate is divided by the average rate, as follows:

$$\frac{400}{250} = \frac{8}{5}$$

Thus, for every 5 words read by the average reader, this man reads 8. Another way of making this comparison is to say that he reads $1\frac{3}{5}$ times as fast as the average reader.

When the relationship between two numbers is shown in this way, they are compared as a ratio. A ratio is a comparison of two like quantities. It is the quotient obtained by dividing the first number of a comparison by the second.

Closely allied with the study of ratio is the subject of proportion. **Proportion** is nothing more than an equation in which the members are ratios. In other words, when two ratios are set equal to each other, a proportion is formed. The proportion may be written in three different ways as in the following examples:

$$15 : 20 :: 3 : 4$$

$$15 : 20 = 3 : 4$$

$$\frac{15}{20} = \frac{3}{4}$$

The last two forms are the most common. All these forms are read, "15 is to 20 as 3 is to 4." In other words, 15 has the same ratio to 20 as 3 has to 4.

One reason for the extreme importance of proportions is that if any three of the terms are given, the fourth may be found by solving a simple equation. In science, many chemical and physical relations are expressed as proportions. Consequently, a familiarity with proportions will provide one method for solving many applied problems. It is evident from the last form shown, $\frac{15}{20} = \frac{3}{4}$, that a proportion is really a fractional equation. Therefore, all the rules for fractional equations apply.

The laws of proportion are listed:

If $a/b = c/d$, then

    (A) $ad = bc$

    (B) $b/a = d/c$

    (C) $a/c = b/d$

    (D) $(a + b)/b = (c + d)/d$

    (E) $(a - b)/b = (c - d)/d$

Given a proportion $a : b = c : d$, then a and d are called extremes, b and c are called the means and d is called the fourth proportion to a, b, and c.

4.1 Ratios

 Which statement below is an incorrect method of expressing a ratio?

a) x to y

b) x:y

c) x = y

d) x/y

 The correct answer is statement c. Statements a and b correctly show how to express a ratio. Statement d is also correct because a ratio can be expressed as a fraction. A ratio cannot be expressed as in statement c.

 Express the following phrases as a ratio of students in Mrs. Polly's class to students in Mr. Smith's:

Mrs. Polly teaches 30 students.

Mr. Smith teaches 10 students.

 Step 1 is to write the phrase in the following format:

30 students:10 students

Step 2 is to reduce the ratio to its lowest terms, if possible. To do this, write out each number as a product of its primes and find the greatest common divisor (GCD).

$$30 = 5 \times 6 \qquad\qquad 10 = 5 \times 2$$

$$= 5 \times 3 \times 2$$

The GCD is 10.

Next, divide each number by the GCD.

$$\frac{30}{10} = 3 \qquad\qquad \frac{10}{10} = 1$$

Step 3 is to rewrite the ratio.

3 students:1 student

The correct answer is 3 students:1 student.

 Express the following phrases as a ratio of cats to dogs:

21 cats are in the house.

7 dogs are in the house.

 Step 1 is to write the phrase in the following format:

21 cats:7 dogs

Step 2 is to reduce the ratio to its lowest terms, if possible. To do this, write out each number as a product of its primes and find the greatest common divisor (GCD).

$$21 = 7 \times 3 \qquad\qquad 7 = 7 \times 1$$

The GCD is 7.

Next, divide each number by the GCD.

$$\frac{21}{7} = 3 \qquad\qquad \frac{7}{7} = 1$$

Step 3 is to rewrite the ratio.

3 cats:1 dog

The correct answer is 3 cats:1 dog.

Express the following phrase as a ratio of dollars lost to dollars invested:

Stockbroker Brian lost $3 for every $6 invested.

 Step 1 is write the phrase in the following format:

−$3:$6

Step 2 is to reduce the ratio to its lowest terms, if possible. To do this, write out each number as a product of its primes and find the greatest common divisor (GCD).

$3 = 3 \times 1$ $6 = 3 \times 2$

The GCD is 3.

Next, divide each number by the GCD.

$\dfrac{3}{3} = 1$ $\dfrac{6}{3} = 2$

Step 3 is to rewrite the ratio. Remember to carry the negative sign into the ratio.

−$1:$2

The correct answer is −$1:$2

4.2 Ratios and Fractions

 Express the following fraction as a ratio: $\dfrac{1}{3}$.

 Step 1 is to rewrite the fraction in the following format:

$\dfrac{1}{3} = 1:3$

Step 2 is to simplify, if possible. To do this, write out each number as a product of its primes and find the greatest common divisor (GCD).

$$1 = 1 \times 1 \qquad\qquad 3 = 3 \times 1$$

Since there is no GCD, the ratio is already reduced to its lowest terms.

The correct answer is 1:3.

 Express the following fraction as a ratio: $\dfrac{54}{63}$.

 Step 1 is to rewrite the fraction in the following format:

$$\frac{54}{63} = 54{:}63$$

Step 2 is to simplify, if possible. To do this, write out each number as a product of its primes and find the greatest common divisor (GCD).

$$54 = 9 \times 6 \qquad\qquad 63 = 9 \times 7$$
$$ = 3 \times 3 \times 3 \times 2 \qquad\qquad = 3 \times 3 \times 7$$

The GCD is 9.

Next, divide each number by the GCD.

$$\frac{54}{9} = 6 \qquad\qquad \frac{63}{9} = 7$$

Step 3 is to rewrite the ratio.

6:7

The correct answer is 6:7.

Express the following fraction as a ratio: $-\dfrac{15}{35}$.

Step 1 is to rewrite the fraction in the following format:

$$-\frac{15}{35} = -15:35$$

Step 2 is to simplify, if possible. To do this, write out each number as a product of its primes and find the greatest common divisor (GCD).

$$15 = 5 \times 3 \qquad\qquad 35 = 5 \times 7$$

The GCD is 5.

Next, divide each number by the GCD.

$$-\frac{15}{5} = -3 \qquad\qquad \frac{35}{5} = 7$$

Step 3 is to rewrite the ratio.

$$-3:7$$

The correct answer is –3:7.

Express the following ratio as a fraction: 5:1.

Step 1 is to determine the numerator and denominator.

The left side of the ratio becomes the numerator: 5.

The right side of the ratio becomes the denominator: 1.

Step 2 is to rewrite the ratio as a fraction.

$$5:1 = \frac{5}{1}$$

Step 3 is to simplify, if possible. To do this, write out each number as a product of its primes and find the greatest common divisor (GCD).

$$5 = 5 \times 1 \qquad\qquad 1 = 1 \times 1$$

Since there is no GCD, the ratio is already reduced to its lowest terms.

Step 4 is to rewrite the fraction.

$$\frac{5}{1}$$

The correct answer is $\frac{5}{1}$.

 Express the following ratio as a fraction: 60:12.

 Step 1 is to determine the numerator and denominator.

The left side of the ratio becomes the numerator: 60.

The right side of the ratio becomes the denominator: 12.

Step 2 is to rewrite the ratio as a fraction.

$$60:12 = \frac{60}{12}$$

Step 3 is to simplify, if possible. To do this, write out each number as a product of its primes and find the greatest common divisor (GCD).

$$60 = 6 \times 10 \qquad\qquad 12 = 6 \times 2$$

$$= 6 \times 5 \times 2 \qquad\qquad = 3 \times 2 \times 2$$

$$= 3 \times 2 \times 5 \times 2$$

The GCD is 12.

Next, divide each number by the GCD.

$$\frac{60}{12} = 5 \qquad\qquad\qquad \frac{12}{12} = 1$$

Step 4 is to rewrite the fraction.

$$\frac{5}{1}$$

The correct answer is $\frac{5}{1}$.

4.3 Proportions

Label the terms in the proportion below.

$$\frac{A}{C} = \frac{B}{D}$$

Terms A and D are known as the extremes. Terms B and C are known as the means. A proportion is defined as two equal ratios where the product of the extremes equals the product of the means, $AD = BC$.

Proportions

 Which of the statements are correct proportions?

a) $\dfrac{4}{1} = \dfrac{1}{4}$

b) $\dfrac{1}{3} = \dfrac{2}{6}$

c) $\dfrac{5}{10} = \dfrac{1}{2}$

d) $\dfrac{2}{3} = \dfrac{-2}{3}$

A The correct answers are statements b and c.

Statement a is incorrect because the product of the extremes does not equal the product of the means.

$AD = BC$ 4(4) does not equal 1(1)

Statement b is correct because the product of the extremes equals the product of the means.

$AD = BC$ 1(6) = 2(3)

Statement c is correct because the product of the extremes equals the product of the means.

$AD = BC$ 5(2) = 1(10)

Statement d is incorrect because the product of the extremes does not equal the product of the means.

$AD = BC$ 2(3) does not equal −2(3)

Find the solution to the proportion below.

$$\frac{3}{7} = \frac{?}{21}$$

 Step 1 is to put the proportion in the following format:

$AD = BC$ $\qquad\qquad\qquad 3(21) = ?(7)$

Step 2 is to solve the left side of the proportion.

$3(21) = 63$

Step 3 is to rewrite the proportion.

$63 = ?(7)$

Step 4 is to find the missing integer that solves the proportion. To do this, divide both sides by the known mean, 7.

$$\frac{63}{7} = 9 \qquad\qquad\qquad \frac{?(7)}{7} = ?$$

Step 5 is to rewrite the proportion.

$9 = ?$

The solution is 9.

Step 6 is to check the answer.

$$\frac{3}{7} = \frac{?}{21}$$

Substituting 9 for ?,

$$\frac{3}{7} = \frac{9}{21}$$

Next, put the proportion in the following format:

$AD = BC$ $3(21) = 9(7)$

Solve both sides of the proportion.

$3(21) = 63$ $9(7) = 63$

Solution "9" checks out correctly.

 Find the solution to the proportion below.

$$\frac{?}{9} = \frac{4}{6}$$

 Step 1 is to put the proportion in the following format:

$AD = BC$ $?(6) = 4(9)$

Step 2 is to solve the right side of the proportion.

$4(9) = 36$

Step 3 is to rewrite the proportion.

$?(6) = 36$

Step 4 is to find the missing integer that solves the proportion. To do this, divide both sides by the known extreme, 6.

$$\frac{?}{6} = ?$$ $$\frac{36}{6} = 6$$

Step 5 is to rewrite the proportion.

$? = 6$

The solution is 6.

Step 6 is to check the answer.

$$\frac{?}{9} = \frac{4}{6}$$

Substituting 6 for ?,

$$\frac{6}{9} = \frac{4}{6}$$

Next, put the proportion in the following format:

$AD = BC$ $\qquad\qquad$ $6(6) = 4(9)$

Solve both sides of the proportion.

$6(6) = 36$ $\qquad\qquad$ $4(9) = 36$

Solution "36" checks out correctly.

 Find the solution to the proportion below.

$$\frac{?}{45} = \frac{-2}{5}$$

 Step 1 is to put the proportion in the following format:

$AD = BC$ $\qquad\qquad$ $?(5) = -2(45)$

Step 2 is to solve the right side of the proportion.

$-2(45) = -90$

Step 3 is to rewrite the proportion.

$?(5) = -90$

Step 4 is to find the missing integer that solves the proportion. To do this, divide both sides by the known extreme, 5.

$$\frac{?}{5} = ?$$ $\qquad\qquad\qquad$ $-\frac{90}{5} = -18$

Step 5 is to rewrite the proportion.

? = −18

The solution is −18.

Step 6 is to check the answer.

$$\frac{?}{45} = -\frac{2}{5}$$

Substituting −18 for ?,

$$-\frac{18}{45} = -\frac{2}{5}$$

Next, put the proportion in the following format.

$AD = BC$ $-18(5) = -2(45)$

Solve both sides of the proportion.

$-18(5) = -90$ $-2(45) = -90$

Solution "−18" checks out correctly.

4.4   Problem Solving with Proportions

A chemist is preparing a chemical solution. She needs to add 3 parts sodium and 2 parts zinc to a flask of chlorine. If she has already placed 300 grams of sodium into the flask, how much zinc must she now add?

Step 1 is to determine the ratio of sodium and zinc.

3 parts sodium, 2 parts zinc = 3:2

Step 2 is to write the problem as a proportion.

$$\frac{3}{2} = \frac{300}{?}$$

Step 3 is to put the proportion in the following format:

$AD = BC$ $3(?) = 2(300)$

Step 4 is to solve the right side of the proportion.

$2(300) = 600$

Step 5 is to rewrite the proportion.

$3(?) = 600$

Step 6 is to find the missing integer that solves the proportion. To do this, divide both sides by the known extreme, 3.

$$\frac{3(?)}{3} = ?$$ $$\frac{600}{3} = 200$$

Step 7 is to rewrite the proportion.

$? = 200$

The solution is 200 grams of zinc.

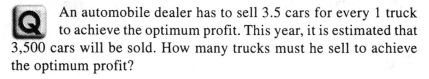 An automobile dealer has to sell 3.5 cars for every 1 truck to achieve the optimum profit. This year, it is estimated that 3,500 cars will be sold. How many trucks must he sell to achieve the optimum profit?

 Step 1 is to determine the ratio of cars to trucks.

3.5 cars, 1 truck = 3.5:1

Make both sides of the ratio an integer. To do this, multiply both sides of the ratio by 2.

$2(3.5):2(1) = 7:2$

Step 2 is to write the problem as a proportion.

$$\frac{7}{2} = \frac{3,500}{?}$$

Step 3 is to put the proportion in the following format:

$AD = BC$　　　　　　　　$7(?) = 2(3,500)$

Step 4 is to solve the right side of the proportion.

$2(3,500) = 7,000$

Step 5 is to rewrite the proportion.

$7(?) = 7,000$

Step 6 is to find the missing integer that solves the proportion. To do this, divide both sides by the known extreme, 7.

$$\frac{7(?)}{7} = ?$$　　　　　　　　$$\frac{7,000}{7} = 1,000$$

Step 7 is to rewrite the proportion.

$? = 1,000$

The solution is 1,000 trucks.

Q A baker is making a new recipe for chocolate chip cookies. He decides that for every 6 cups of flour, he needs to add 1 cup of sugar. He puts 30 cups of flour and 2 cups of sugar into the batter. How much more sugar does he need?

 A Step 1 is to determine the ratio of flour to sugar.

6 cups flour, 1 cup sugar = 6:1

Step 2 is to write the problem as a proportion.

$$\frac{6}{1} = \frac{30}{?}$$

Step 3 is to put the proportion in the following format:

$AD = BC$ $6(?) = 1(30)$

Step 4 is to solve the right side of the proportion.

$1(30) = 30$

Step 5 is to rewrite the proportion.

$6(?) = 30$

Step 6 is to find the missing integer that solves the proportion. To do this, divide both sides by the known extreme, 6.

$$\frac{6(?)}{6} = ? \qquad\qquad \frac{30}{6} = 5$$

Step 7 is to rewrite the proportion.

$? = 5$

The solution is that 5 cups of sugar must be added to the batter.

Step 8 is to determine how many more cups of sugar are needed.

$5 - 2 = 3$

Since only 2 cups have been added so far, the baker must still add 3 cups.

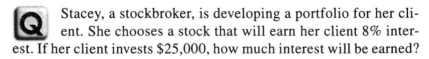

Stacey, a stockbroker, is developing a portfolio for her client. She chooses a stock that will earn her client 8% interest. If her client invests $25,000, how much interest will be earned?

Step 1 is to determine the ratio of interest earned per dollar invested. Remember that fractions can also express ratios.

$$8\% = \frac{8}{100}$$

$$\frac{8}{100} = 8{:}100$$

Step 2 is to divide both sides of the ratio by the GCD to simplify the expression.

$$8 = 2 \times 4 \qquad\qquad 100 = 10 \times 10$$

$$= 2 \times 2 \times 2 \qquad\qquad = 5 \times 2 \times 5 \times 2$$

The GCD is 4; divide both sides of the proportion by 4.

$$\frac{8}{4} = 2 \qquad\qquad \frac{100}{4} = 25$$

The new proportion is 2:25.

Step 3 is to write the problem as a proportion.

$$\frac{2}{25} = \frac{?}{25{,}000}$$

Step 4 is to put the proportion in the following format:

$$AD = BC \qquad\qquad 2(25{,}000) = ?(25)$$

Step 5 is to solve the left side of the proportion.

$$2(25{,}000) = 50{,}000$$

Step 6 is to rewrite the proportion.

$$50{,}000 = ?(25)$$

Step 7 is to find the missing integer that solves the proportion. To do this, divide both sides by the known mean, 25.

$$\frac{50,000}{25} = 2,000 \qquad\qquad \frac{?(25)}{25} = ?$$

Step 8 is to rewrite the proportion.

2,000 = ?

The solution is that $2,000 of interest will be earned.

 Elena works as a traveling salesperson for Knives Incorporated. Each year, she increases the number of customers proportionally (5:1). If in her first year she started with 100 customers, how many does she have by the end of the second year?

Step 1 is to write the problem as a proportion.

$$\frac{5}{1} = \frac{?}{100}$$

Step 2 is to put the proportion in the following format:

$AD = BC$ $\qquad\qquad$ 5(100) = 1(?)

Step 3 is to solve the left side of the proportion.

5(100) = 500

Step 4 is to rewrite the proportion.

500 = 1(?)

Step 5 is to add the number of new customers to the original number of customers.

100 + 500 = 600

Therefore, after her first year, she has 600 customers. Repeat the above steps for the second year.

Step 6 is to rewrite the problem as a proportion.

$$\frac{5}{1} = \frac{?}{600}$$

Step 7 is to put the proportion in the following format:

$AD = BC$ $5(600) = 1(?)$

Step 8 is to solve the left side of the proportion.

$5(600) = 3,000$

Step 9 is to rewrite the proportion.

$3,000 = (1)?$

Step 10 is to find the missing integer that solves the proportion. To do this, divide both sides by the known mean, 1.

$$\frac{3,000}{1} = 3,000 \qquad\qquad \frac{(1)?}{1} = ?$$

Step 11 is to rewrite the proportion.

$3,000 = ?$

Step 12 is to add the number of new customers to the number of customers acquired during the first year.

$600 + 3,000 = 3,600$

The correct answer is 3,600 customers.

CHAPTER 5

Percents

The word "percent" is derived from Latin. It was originally "per centum," which means "by the hundred." Thus the statement is often made that "percent means hundredths."

Percentage deals with the group of decimal fractions whose denominators are 100—that is, fractions of two decimal places. Since hundredths were used so frequently, the decimal point was dropped and the symbol % was placed after the number and read "percent" (per 100). Thus, 0.15 and 15% represent the same value, 15/100. The first is read "15 hundredths," and the second is read "15 percent." Both mean 15 parts out of 100.

Thus, a **percent** is a way of expressing the relationship between part and whole, where whole is defined as 100%. A percent can be defined by a fraction with a denominator of 100. Decimals can also represent a percent. For instance,

$$56\% = 0.56 = 56/100$$

5.1   Percents and Fractions

 Convert the following fraction to a percent: $\dfrac{1}{2}$.

A   Percents are based on fractions with a denominator of 100. Therefore, $\frac{1}{2}$ has to be converted to an equivalent fraction with a denominator of 100.

Step 1 is to write the problem as a proportion.

$$\frac{1}{2} = \frac{?}{100}$$

Step 2 is to rewrite the proportion in the following format:

$AD = BC$ $1(100) = ?(2)$

Step 3 is to solve the left side of the proportion.

$1(100) = 100$

Step 4 is to rewrite the proportion.

$100 = ?(2)$

Step 5 is to find the missing integer that solves the proportion. To do this, divide both sides by the known mean, 2.

$$\frac{100}{2} = 50 \qquad\qquad \frac{?(2)}{2} = ?$$

Step 6 is to rewrite the proportion.

$50 = ?$

The new fraction is $\frac{50}{100}$.

Step 7 is to change the numerator to a percent.

$50 = 50\%$

The answer is 50%.

Convert the following fraction to a percent: $\frac{17}{20}$.

 Step 1 is to write the problem as a proportion.

$$\frac{17}{20} = \frac{?}{100}$$

Step 2 is to rewrite the proportion in the following format:

$AD = BC$ $\qquad\qquad$ $17(100) = ?(20)$

Step 3 is to solve the left side of the proportion.

$17(100) = 1,700$

Step 4 is to rewrite the proportion.

$1,700 = ?(20)$

Step 5 is to find the missing integer that solves the proportion. To do this, divide both sides by the known mean, 20.

$$\frac{1,700}{20} = 85 \qquad\qquad \frac{?(20)}{20} = ?$$

Step 6 is to rewrite the proportion.

$85 = ?$

The new fraction is $\frac{85}{100}$.

Step 7 is to change the numerator to a percent.

$85 = 85\%$

The answer is 85%.

Q Convert the following fraction to a percent: $\dfrac{565}{1,000}$.

A Since this fraction has a denominator of 1000, the percent will have a digit in the tenths' place.

Step 1 is to write the problem as a proportion.

$$\frac{565}{1,000} = \frac{?}{100}$$

Step 2 is to rewrite the proportion in the following format:

$AD = BC$ $565(100) = ?(1,000)$

Step 3 is to solve the left side of the proportion.

$565(100) = 56,500$

Step 4 is to rewrite the proportion.

$56,500 = ?(1,000)$

Step 5 is to find the missing integer that solves the proportion. To do this, divide both sides by the known mean, 1,000.

$$\frac{56,500}{1,000} = 56.5 \qquad\qquad \frac{?(1,000)}{1,000} = ?$$

Step 6 is to rewrite the proportion.

$56.5 = ?$

The new fraction is $\dfrac{56.5}{100}$.

Step 7 is to change the numerator to a percent.

$56.5 = 56.5\%$

The answer is 56.5%.

 Convert the following fraction into a percent: $-\dfrac{8}{25}$.

 The negative sign will not affect the problem, but it must be carried through.

Step 1 is to write the problem as a proportion.

$$-\frac{8}{25} = \frac{?}{100}$$

Step 2 is to rewrite the proportion in the following format:

$AD = BC$ $-8(100) = ?(25)$

Step 3 is to solve the left side of the proportion.

$-8(100) = -800$

Step 4 is to rewrite the proportion.

$-800 = ?(25)$

Step 5 is to find the missing integer that solves the proportion. To do this, divide both sides by the known mean, 25.

$$-\frac{800}{25} = -32 \qquad\qquad \frac{?(25)}{25} = ?$$

Step 6 is to rewrite the proportion.

$-32 = ?$

The new fraction is $\dfrac{-32}{100}$.

Step 7 is to change the numerator to a percent.

$$-32 = -32\%$$

The answer is –32%.

Convert the following percent to a fraction: 68%.

Step 1 is to convert 68% into a fraction with a denominator of 100.

$$68\% = \frac{68}{100}$$

Step 2 is to simplify the fraction.

$$\frac{68}{100} = \frac{17}{25}$$

The correct answer is $\frac{17}{25}$.

Convert the following percent to a fraction: 24.6%.

Step 1 is to convert 24.6% into a fraction with a denominator of 1,000. The denominator cannot be 100 in this problem because the percent has a digit in the tenths' place. So, in order to convert this percent into a fraction, a denominator of 1,000 must be used.

$$24.6\% = \frac{246}{1,000}$$

Step 2 is to simplify the fraction.

$$\frac{246}{1,000} = \frac{123}{500}$$

The correct answer is $\frac{123}{500}$.

 Convert the following percent into a fraction: –45%.

 Step 1 is to convert –45% into a fraction with a denominator of 100. The negative will have no effect on the problem, but it must be carried through.

$$-45\% = -\frac{45}{100}$$

Step 2 is to simplify the fraction.

$$-\frac{45}{100} = -\frac{9}{20}$$

The correct answer is $-\frac{9}{20}$.

 Solve the following operation. Convert the answer to a percent.

$$2\frac{1}{5} + \frac{1}{10}$$

 Step 1 is to find the least common denominator for $2\frac{1}{5}$ and $\frac{1}{10}$.

$$2\frac{1}{5} = \frac{22}{10} \qquad\qquad \frac{1}{10} = \frac{1}{10}$$

Step 2 is to perform the operation.

$$\frac{22}{10} + \frac{1}{10} = \frac{23}{10}$$

Step 3 is to write the problem as a proportion.

$$\frac{23}{10} = \frac{?}{100}$$

Step 4 is to rewrite the proportion in the following format:

$$AD = BC \qquad\qquad 23(100) = ?(10)$$

Step 5 is to solve the left side of the proportion.

$$23(100) = 2,300$$

Step 6 is to rewrite the proportion.

$$2,300 = ?(10)$$

Step 7 is to find the missing integer that solves the proportion. To do this, divide both sides by the known mean, 10.

$$\frac{2,300}{10} = 230 \qquad\qquad \frac{?(10)}{10} = ?$$

Step 8 is to rewrite the proportion.

$$230 = ?$$

The new fraction is $\frac{230}{100}$.

Step 9 is to change the numerator to a percent.

$$230 = 230\%$$

The answer is 230%.

 Solve the following operation. Convert the answer to a fraction.

$$25\% + 8\%$$

 Step 1 is to perform the operation.

$$25\% + 8\% = 33\%$$

Step 2 is to convert 33% into a fraction with a denominator of 100.

$$33\% = \frac{33}{100}$$

Step 3 is to simplify the fraction.

$$\frac{33}{100} = \frac{33}{100}$$

Since $\frac{33}{100}$ cannot be simplified, the correct answer is $\frac{33}{100}$.

5.2 Percents and Decimals

 Convert the following decimal to a percent: 0.77.

 Percents are based on fractions with a denominator of 100 or decimals carried to the hundredths' place.

Step 1 is to multiply the decimal by 100.

0.77 × 100 = 77.00

Step 2 is to write the decimal as a percent.

77.00 = 77%

The correct answer is 77%.

 Convert the following decimal to a percent: 0.02.

 Step 1 is to multiply the decimal by 100.

0.02 × 100 = 2.00

Step 2 is to write the decimal as a percent.

2.00 = 2%

The correct answer is 2%.

 Convert the following decimal to a percent: 12.69.

 Step 1 is to multiply the decimal by 100.

12.69 × 100 = 1,269.00

Step 2 is to write the decimal as a percent.

1,269.00 = 1,269%

The correct answer is 1,269%.

Convert the following decimal to a percent: 0.0094.

 This decimal has digits in places less than the tenths' or hundredths' place. However, the steps for converting to a percent are the same.

Step 1 is to multiply the decimal by 100.

$0.0094 \times 100 = 0.940$

Step 2 is to write the decimal as a percent.

$0.940 = 0.94\%$

The correct answer is 0.94%.

 Convert the following decimal to a percent: −0.26.

 The negative sign will have no effect on the problem. The steps for converting to a percent are the same, but the negative sign must be carried through the problem.

Step 1 is to multiply the decimal by 100.

$-0.26 \times 100 = -26.00$

Step 2 is to write the decimal as a percent.

$-26.00 = -26\%$

The correct answer is −26%.

Convert the following percent to a decimal: 8%.

Step 1 is to write the percent as a real number.

$8\% = 8$

Step 2 is to divide the real number by 100.

$8 \div 100 = 0.08$

The correct answer is 0.08.

 Convert the following percent to a decimal: 233%.

 Although 233% is greater than 100%, the steps for converting to a decimal remain the same.

Step 1 is to write the percent as a real number.

$233\% = 233$

Step 2 is to divide the real number by 100.

$233 \div 100 = 2.33$

The correct answer is 2.33.

 Convert the following percent to a decimal: –10%.

 The negative sign will have no effect on the problem. The steps for converting to a percent are the same, but the negative sign must be carried through the problem.

Step 1 is to write the percent as a real number.

$-10\% = -10$

Step 2 is to divide the real number by 100.

$-10 \div 100 = -0.10$

The correct answer is –0.10.

Perform the following operation. Convert the answer to a percent.

0.99 – 0.45

 Step 1 is to perform the operation.

0.99 – 0.45 = 0.54

Step 2 is to multiply the decimal by 100.

0.54 × 100 = 54.00

Step 3 is to write the decimal as a percent.

54.00 = 54%

The correct answer is 54%.

 Perform the following operation. Convert the answer to a decimal.

145% + 2%

 Step 1 is to perform the operation.

145% + 2% = 147%

Step 2 is to write the percent as a real number.

147% = 147

Step 3 is to divide the real number by 100.

147 ÷ 100 = 1.47

The correct answer is 1.47.

Which of the following statements below are true?

a) 0.002% = 0.200

b) 1.967 = 196.7%

c) −0.95 > −93%

d) 1.00 < 100%

 Statement a is incorrect because 0.200 is equivalent to 20%. Since 20% does not equal 0.002%, this statement cannot be true.

Statement b is correct. 1.967 is equivalent to 196.7%.

Statement c is incorrect because −0.95 is equivalent to −95%. Since −95% is not greater than −93%, this statement is not true.

Statement d is incorrect because 1.00 is equivalent to 100%, not less than 100%.

The only correct statement is statement b.

5.3   Problem Solving with Percents

 What is 45% of 100?

 This problem has to be made into a multiplication operation.

Step 1 is to convert 45% into a decimal.

45% = 45 ÷ 100 = 0.45

Step 2 is to set up the multiplication operation.

100 × 0.45

Step 3 is to perform the operation.

$$100 \times 0.45 = 45$$

The correct answer is 45.

 What is 10% of –21?

 This problem has to be made into a multiplication operation.

Step 1 is to convert 10% into a decimal.

$$10\% = 10 \div 100 = 0.10$$

Step 2 is to set up the multiplication operation.

$$-21 \times 0.10$$

Step 3 is to perform the operation.

$$-21 \times 0.10 = -2.10$$

The correct answer is –2.10.

 What is 200% of 5?

 This problem has to be made into a multiplication operation.

Step 1 is to convert 200% into a decimal.

$$200\% = 200 \div 100 = 2.00$$

Step 2 is to set up the multiplication operation.

$$5 \times 2.00$$

Step 3 is to perform the operation.

$5 \times 2.00 = 10$

The correct answer is 10.

 70 is what percentage of 80?

 This problem has to be made into a division operation.

Step 1 is to set up the division operation.

$70 \div 80$

Step 2 is to perform the division.

$70 \div 80 = 0.875$

Step 3 is to convert the decimal into a percent.

$0.875 \times 100 = 87.5\%$

The correct answer is 87.5%.

 45 is what percentage of 40?

 This problem has to be made into a division operation.

Step 1 is to set up the division operation.

$45 \div 40$

Step 2 is to perform the division.

$45 \div 40 = 1.125$

Step 3 is to convert the decimal into a percent.

$1.125 \times 100 = 112.5\%$

The correct answer is 112.5%.

 What is 25% of $\dfrac{1}{2}$?

 This problem has to be made into a multiplication operation.

Step 1 is to convert 25% into a decimal.

$25\% = 25 \div 100 = 0.25$

Step 2 is to set up the multiplication operation.

$\dfrac{1}{2} \times 0.25$ or $\dfrac{1}{2} \times \dfrac{1}{4}$

Step 3 is to perform the operation.

$\dfrac{1}{2} \times \dfrac{1}{4} = \dfrac{1}{8}$

The correct answer is $\dfrac{1}{8}$.

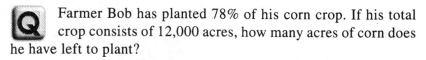 Farmer Bob has planted 78% of his corn crop. If his total crop consists of 12,000 acres, how many acres of corn does he have left to plant?

 This problem has to be made into a multiplication operation.

Step 1 is to convert 78% into a decimal.

$78\% = 78 \div 100 = 0.78$

Step 2 is to set up the multiplication operation.

   12,000 acres × 0.78

Step 3 is to perform the operation.

   12,000 acres × 0.78 = 9,360 acres

Step 4 is to subtract 9,360 from 12,000 to find the number of acres left.

   12,000 − 9,360 = 2,640 acres

The correct answer is 2,640 acres.

 Ms. Harper is a used car salesperson. Every month she is required to sell 20 cars. This month, however, she has sold 110% of the monthly requirement. How many cars did she sell?

 This problem has to be made into a multiplication operation.

Step 1 is to convert 110% into a decimal.

   110% = 110 ÷ 100 = 1.10

Step 2 is to set up the multiplication operation.

   20 × 1.10

Step 3 is to perform the operation.

   20 × 1.10 = 22

The correct answer is 22 cars.

 State University has a very competitive engineering program. After 1 week, 40% of the students drop the program. If 250 students start the program, how many are left after the first week?

This problem has to be made into a multiplication operation.

Step 1 is to convert 40% into a decimal.

$40\% = 40 \div 100 = 0.40$

Step 2 is to set up the multiplication operation.

250×0.40

Step 3 is to perform the operation $250 \times 0.40 = 100$

Therefore, 100 students drop the program after 1 week.

Step 4 is to find how many students remain.

$250 - 100 = 150$

The correct answer is 150 students remain.

 A child is discovered to have a rare but non-fatal virus. Research has found that 0.0005% of people that contact those that are infected with the virus will contract the virus as well. If 8 people were known to have contact with the child, how many people will get the virus?

 This problem has to be made into a multiplication operation.

Step 1 is to convert 0.0005% into a decimal.

$0.0005\% = 0.0005 \div 100 = 0.000005$

Step 2 is to set up the multiplication operation.

8×0.000005

Step 3 is to perform the operation.

$8 \times 0.000005 = 0.00004$

The solution is 0.00004 people. This number is not realistic because 0.00004 people cannot exist. Therefore, we have to round the answer to the nearest whole number.

The correct answer is 0 people.

 Central City has a yearly marathon. Fifty-six people signed up for the marathon last year but only 40 people actually completed. What percent of the people that signed up did not compete in the race? Round the answer to the nearest whole number percent.

This problem has to be made into a division operation.

Step 1 is to set up the division operation.

40 ÷ 56

Step 2 is to perform the division.

40 ÷ 56 = 0.71

Step 3 is to convert the decimal into a percent.

0.71 × 100 = 71%

Therefore, 71% of those that signed up competed in the race.

Step 4: to get the percent that did not compete, subtract 71% from 100%.

100% − 71% = 29%

The correct answer is 29% of those that signed up did not compete in the race.

CHAPTER 6

Exponents and Roots

6.1   Exponents

When a number is multiplied by itself a specific number of times, it is said to be **raised to a power**. The way this is written is $a^n = b$ where a is the number or **base**, n is the **exponent** or **power** that indicates the number of times the base is to be multiplied by itself, and b is the product of this multiplication.

In the expression 3^2, 3 is the base and 2 is the exponent. This means that 3 is multiplied by itself 2 times and the product is 9.

An exponent can be either positive or negative. A negative exponent implies a fraction. Such that, if n is a positive integer

$$a^{-n} = \frac{1}{a^n}, a \neq 0. \text{ So, } 2^{-4} = \frac{1}{2^4} = \frac{1}{16}.$$

An exponent that is zero gives a result of 1, assuming that the base is not equal to zero.

$a^0 = 1, a \neq 0.$

An exponent can also be a fraction. If m and n are positive integers,

$$a^{\frac{m}{n}} = \sqrt[n]{a^m}$$

The numerator remains the exponent of a, but the denominator tells what root to take. For example,

(1) $\quad 4^{\frac{3}{2}} = \sqrt[2]{4^3} = \sqrt{64} = 8$

(2) $\quad 3^{\frac{4}{2}} = \sqrt[2]{3^4} = \sqrt{81} = 9$

If a fractional exponent were negative, the same operation would take place, but the result would be a fraction. For example,

(1) $\quad 27^{-\frac{2}{3}} = \frac{1}{27^{2/3}} = \frac{1}{\sqrt[3]{27^2}} = \frac{1}{\sqrt[3]{729}} = \frac{1}{9}$

6.2   Radicals

The **square root** of a number is a number that when multiplied by itself results in the original number. So, the square root of 81 is 9 since $9 \times 9 = 81$. However, -9 is also a root of 81 since $(-9)(-9) = 81$. Every positive number will have two roots. Yet, the principal root is the positive one. Zero has only one square root, while negative numbers do not have real numbers as their square roots.

A **radical sign** indicates that the root of a number or expression will be taken. The **radicand** is the number of which the root will be taken. The **index** tells how many times the root needs to be multiplied by itself to equal the radicand. E.g.,

index

radical sign → √ radicand

(1) $\sqrt[3]{64}$;

3 is the index and 64 is the radicand. Since

$4 \cdot 4 \cdot 4 = 64$, $\sqrt[3]{64} = 4$

(2) $\sqrt[5]{32}$;

5 is the index and 32 is the radicand. Since

$2 \times 2 \times 2 \times 2 \times 2 = 32$, $\sqrt[5]{32} = 2$

6.3   Powers and Roots

 Find the solution to the following problem: 2^3.

 Step 1 is to identify the base and the exponent. In this problem, "2" is the base and "3" is the exponent.

Step 2 is to set up the multiplication. Multiply the base, "2," with itself "3" times.

$2 \times 2 \times 2$

Step 3 is to perform the operation.

$2 \times 2 \times 2 = 8$

The correct answer is 8.

 Find the solution to the following problem: $(-8)^2$.

 Step 1 is to identify the base and the exponent. In this problem, "–8" is the base and "2" is the exponent.

Step 2 is to set up the multiplication. Multiply the base, "–8," with itself "2" times.

$$-8 \times -8$$

Step 3 is to perform the operation. Since a negative base becomes a positive number when multiplied with an even-numbered positive exponent, the result will be positive.

$$-8 \times -8 = 64$$

The correct answer is 64.

 Find the solution to the following problem: $(-3)^5$.

 Step 1 is to identify the base and the exponent. In this problem, "–3" is the base and "5" is the exponent.

Step 2 is to set up the multiplication. Multiply the base, "–3," with itself "5" times.

$$-3 \times -3 \times -3 \times -3 \times -3$$

Step 3 is to perform the operation. Since a negative base becomes a negative number when multiplied with an odd-numbered exponent, the result will be negative.

$$-3 \quad -3 \quad -3 \quad -3 \quad -3 = -243$$

The correct answer is –243.

 Find the solution to the following problem: 4^1.

 Step 1 is to identify the base and the exponent. In this problem, "4" is the base and "1" is the exponent.

Step 2 is to set up the multiplication. Multiply the base, "4," with itself "1" time. The result of any base raised to "1" will always be the base.

The correct answer is 4.

 Find the solution to the following problem: 25^0.

 Step 1 is to identify the base and the exponent. In this problem, "25" is the base and "0" is the exponent.

Step 2 is to set up the multiplication. Multiply the base, "25," with itself "0" times. The result of any base raised to "0" will always be 1.

The correct answer is 1.

 Find the solution to the following problem: 4^{-2}.

 Step 1 is to identify the base and the exponent. In this problem, "4" is the base and "–2" is the exponent.

Step 2 is to set up the multiplication. Multiply the base, "4," with itself "2" times.

$$4 \times 4$$

Step 3 is to perform the operation.

$$4 \times 4 = 16$$

Step 4: since the exponent is negative, the problem becomes inverted.

$$16 \text{ becomes } \frac{1}{16}$$

The correct answer is $\frac{1}{16}$.

 Find the solution to the following problem: $(5^2)(2^3)$.

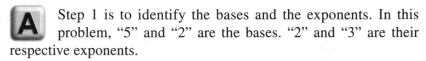 Step 1 is to identify the bases and the exponents. In this problem, "5" and "2" are the bases. "2" and "3" are their respective exponents.

Step 2: since the problem does not contain a common base, the exponents cannot be added. Instead, set up the multiplication for each base separately.

$$5^2 = 5 \times 5$$

$$2^3 = 2 \times 2 \times 2$$

Step 3 is to perform each operation.

$$5 \times 5 = 25$$

$$2 \times 2 \times 2 = 8$$

Step 4 is to multiply the results together.

$$25 \times 8 = 200$$

The correct answer is 200.

 Find the solution to the following problem: $(-6)^2 (7^2)$.

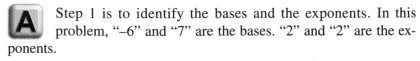 Step 1 is to identify the bases and the exponents. In this problem, "–6" and "7" are the bases. "2" and "2" are the exponents.

Step 2: since the problem does not contain a common base, the exponents cannot be added. Instead, set up the multiplication for each base separately.

$$(-6)^2 = -6 \times -6$$

$$7^2 = 7 \times 7$$

Step 3 is to perform each operation.

$$-6 \times -6 = 36$$

$$7 \times 7 = 49$$

Step 4 is to multiply the results together.

$$36 \times 49 = 1,764$$

The correct answer is 1,764.

Note: -6^2 means $-(6)(6) = -36$

 Find the solution to the following problem: $\dfrac{\left(6^6\right)}{\left(6^4\right)}$.

 Step 1 is to identify the base and the exponents. In this problem, "6" is the common base. "6" and "4" are the exponents.

Step 2: since the problem contains a common base, the exponents can be subtracted.

$$6 - 4 = 2$$

Step 3 is to rewrite the problem using the new exponent.

$$6^2$$

Step 4 is to perform the operation.

$$6 \times 6 = 36$$

The correct answer is 36.

Find the solution to the following problem: $(7^{-3}) \div (7^{-5})$.

 Step 1 is to identify the bases and the exponents. In this problem, "7" is the common base. "−3" and "−5" are the exponents.

Step 2: since the problem contains a common base, the exponents can be subtracted.

$$-3 - (-5) = 2$$

Step 3 is to rewrite the problem using the new exponent.

7^2

Step 4 is to perform the operation.

$$7 \times 7 = 49$$

The correct answer is 49.

 Find the solution to the following problem: $\dfrac{(4^3)}{(2^{-1})}$.

 Step 1 is to identify the bases and the exponents. In this problem, "4" and "2" are the bases. "3" and "−1" are the exponents.

Step 2: since the problem does not contain a common base, the exponents cannot be subtracted. Instead, set up the multiplication for each base separately.

$$4^3 = 4 \times 4 \times 4 \qquad\qquad 2^{-1} = \frac{1}{2^1}$$

Step 3 is to perform each operation.

$$4 \times 4 \times 4 = 64 \qquad \frac{1}{2^1} = \frac{1}{2}$$

Step 4 is to divide the results.

$$64 \div \frac{1}{2} = 128$$

The correct answer is 128.

Find the solution to the following problem: $\sqrt{49}$.

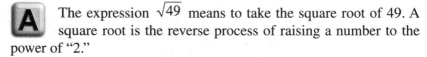

The expression $\sqrt{49}$ means to take the square root of 49. A square root is the reverse process of raising a number to the power of "2."

Step 1 is to determine what the base would be if you raise a number to the power of "2" to get 49.

Base = 7

Step 2 is to raise the base "7" to the exponent "2" to verify the solution.

$$7 \times 7 = 49$$

The correct answer is 7.

Find the solution to the following problem: $\sqrt[3]{8}$.

The expression $\sqrt[3]{8}$ means to take the cube root of 8. A cube root is the reverse process of raising a number to the power of "3."

Step 1 is to determine what the base would be if you raise a number to the power of "3" to get 8.

Base = 2

Step 2 is to raise the base "2" to the exponent "3" to verify the solution.

$2 \times 2 \times 2 = 8$

The correct answer is 2.

 Find the solution to the following problem: $\sqrt[4]{81}$.

 The expression $\sqrt[4]{81}$ means to take the 4th root of 81. A 4th root is the reverse process of raising a number to the power of "4."

Step 1 is to determine what the base would be if you raise a number to the power of "4" to get 81.

Base = 3

Step 2 is to raise the base "3" to the exponent "4" to verify the solution.

$3 \times 3 \times 3 \times 3 = 81$

The correct answer is 3.

 Find the solution to the following problem: $\sqrt[3]{27}$.

 Step 1 is to determine what the base would be if you raise a number to the power of "3" to get 27.

Base = 3

Step 2 is to raise the base "3" to the exponent "3" to verify the solution.

$3 \times 3 \times 3 = 27$

The correct answer is 3.

 Find the solution to the following problem: $\sqrt[3]{-1}$.

 Step 1 is to determine what the base would be if you raise a number to the power of "3" to get -1.

Base $= -1$

Step 2 is to raise the base "-1" to the exponent "3" to verify the solution.

$-1 \times -1 \times -1 = -1$

The correct answer is -1.

6.4 Irrational Numbers

 Find the solution to the following problem: $\sqrt{-11}$.

 There is no solution to this problem. A positive root of a negative number is not a real number.

 Find the solution to the following problem: $\sqrt{-25}$.

 There is no solution to this problem. A positive root of a negative number is not a real number.

Find the solution to the following problem: $\sqrt{2}$.

Step 1 is to determine what the base would be if you raise a number to the power of "2" to get $\sqrt{2}$.

Base = ?

Since the base is going to be a non-integer, it is best to leave the expression as a root.

The correct answer is $\sqrt{2}$. This is known as an irrational number.

Find the solution to the following problem: $\sqrt{50}$.

Some square roots can be expressed as a product of two or more square roots.

Step 1 is to determine if $\sqrt{50}$ can be expressed as a product of two or more square roots.

$$\sqrt{50} = \sqrt{(2 \times 25)} = \sqrt{2}\sqrt{25}$$

Step 2 is to solve the square root that will not be an irrational number.

$$\sqrt{25} = 5$$

Step 3 is to rewrite the problem.

$$\sqrt{50} = 5\left(\sqrt{2}\right)$$

Step 4 is to determine if the problem is irrational. If so, then stop.

$5\left(\sqrt{2}\right)$ is irrational.

The correct answer is $5\left(\sqrt{2}\right)$.

CHAPTER 7

Introduction to

Algebra

In algebra, letters or variables are used to represent numbers. A **variable** is defined as a placeholder, which can take on any of several values at a given time. A **constant**, on the other hand, is a symbol which takes on only one value at a given time. A **term** is a constant, a variable, or a combination of constants and variables. For example: 7.76, $3x$, xyz, $5z/x$, $(0.99)x^2$ are terms. If a term is a combination of constants and variables, the constant part of the term is referred to as the **coefficient** of the variable. If a variable is written without a coefficient, the coefficient is assumed to be 1.

$3x^2$	y^3
coefficient: 3	coefficient: 1
variable: x	variable: y

An **expression** is a collection of one or more terms. If the number of terms is greater than 1, the expression is said to be the sum of the terms.

$$9, 9xy, 6x + x/3, 8yz - 2x$$

An algebraic expression consisting of only one term is called a **monomial**, of two terms is called a **binomial**, of three terms is called a **trinomial**. In general, an algebraic expression consisting of two or more terms is called a **polynomial**.

7.1   True Equations and Inequalities

 Label the expressions in the following algebraic equation.

$$3x^2 + 5 - 12x^2 + x = 27$$

A   B   CD   E   F

 Expression A is known as a variable. In algebra, a variable represents an unknown value. Expression B is known as a constant. A constant does not have a variable associated with it. The constant's value will change when the equation is being factored, simplified, or solved. Expression C is known as a coefficient to the variable x^2. This means that whatever the value for x^2 is, it will be multiplied by 12. Expression D is known as a variable. In algebra, a variable represents an unknown value. Expressions A and D are known as like (or similar) terms. Since the variable for $3x^2$ and $12x^2$ is x^2, $3x^2$ and $12x^2$ are called like terms. Expression E is known as a variable. In algebra, a variable represents an unknown value. Notice there is no coefficient for x. When a variable is written without a coefficient, the coefficient is always "1." In this problem, x could be written as $1x$, but the "1" is always omitted. Expression F is known as a constant. A constant does not have a variable associated with it. The constant's value will change when the equation is being factored, simplified, or solved.

 Verify that the algebraic expression below is false.

$$(50 - 45) \times \frac{3}{17} = x, \text{ if } x = -\frac{15}{17}$$

 The left side of the equation must be simplified.

Step 1 is to perform the operation inside the parentheses.

$$50 - 45 = 5$$

Step 2 is to rewrite the equation, substituting $-\dfrac{15}{17}$ for x.

$$5 \times \frac{3}{17} = -\frac{15}{17}$$

Step 3 is to perform the multiplication operation.

$$5 \times \frac{3}{17} = \frac{15}{17}$$

Step 4 is to rewrite the equation.

$$\frac{15}{17} = -\frac{15}{17}$$

Since $\dfrac{15}{17}$ does not equal $-\dfrac{15}{17}$, the equation is false.

Verify that the algebraic expression below is true.

$3x + 12 = 24$, if $x = 4$

The left side of the equation must be solved.

Step 1 is to substitute a value for the variable x. Since $x = 4$, substitute 4 for x in the equation.

$$3(4) + 12 = 24$$

Step 2 is to perform the multiplication operation.

$$3 \times 4 = 12$$

Step 3 is to rewrite the equation.

$$12 + 12 = 24$$

Step 4 is to perform the addition operation.

12 + 12 = 24

Step 5 is to rewrite the equation.

24 = 24

Since 24 = 24, this algebraic equation is true.

 Verify that the algebraic expression below is true.

$$\frac{(3 \times 15)}{9} \geq x, \text{ if } x = 3$$

 The left side of the inequality must be simplified.

Step 1 is to perform the operation inside the parentheses.

3 × 15 = 45

Step 2 is to rewrite the inequality, substituting 3 for *x*.

$$\frac{45}{9} \geq 3$$

Step 3 is to perform the division operation.

$$\frac{45}{9} = 5$$

Step 4 is to rewrite the inequality.

5 ≥ 3.

Since 5 ≥ 3, the inequality is true.

Verify that the algebraic expression below is true.

$$\left(-\frac{7}{2}\right) \div \frac{1}{2}y < 4, \text{ if } y = 7$$

The left side of the inequality must be solved.

Step 1 is to substitute a value for the variable y. Since $y = 7$, substitute 7 for y in the inequality.

$$\left(-\frac{7}{2}\right) \div \frac{1}{2}(7) < 4$$

Step 2 is to perform the multiplication operation.

$$\frac{1}{2} \times 7 = \frac{7}{2}$$

Step 3 is to rewrite the inequality.

$$\left(-\frac{7}{2}\right) \div \frac{7}{2} < 4$$

Step 4 is to perform the division operation.

$$\left(-\frac{7}{2}\right) \div \frac{7}{2} = -1$$

Step 5 is to rewrite the inequality.

$$-1 < 4$$

Since $-1 < 4$, this algebraic expression is true.

7.2   Commutative Property

 Which of the following statements illustrates the Commutative Property of Addition?

a) $12 + 24 = 12 + x$

b) $69 + 3x = 3(x - 23)$

c) $10x + 12y = 12y + 10x$

d) $2(2 + 3) = 10$

 The correct answer is statement c. Statement c correctly illustrates the Commutative Property of Addition. This property states that order is not relevant when performing addition. In statement c, $10x + 12y = 12y + 10x$, both sides of the equation will be equal, showing that order is not relevant in addition. Statements a, b, and d do not illustrate the Commutative Property of Addition.

 Which of the following statements illustrates the Commutative Property of Multiplication?

a) $xy = x + y$

b) $2x(-45y) = -45y(2x)$

c) $35(2) = \dfrac{70}{1}$

d) $5^2 = 25$

 The correct answer is statement b. Statement b correctly illustrates the Commutative Property of Multiplication. This property states that order is not relevant when performing multiplication. In statement b, $2x(-45y) = -45y(2x)$, both sides of the equation will be equal, showing that order is not relevant in multiplication. Statements a, c, and d do not illustrate the Commutative Property of Multiplication.

 Using the Commutative Property of Multiplication, fill in the missing part of the equation below.

$$8\left(\frac{10}{12}\right) = ?$$

A The Commutative Property of Multiplication states that order is not relevant in multiplication. The format for this property is shown below.

$$A(B) = B(A)$$

Step 1 is to determine what terms represent A and B in the given equation.

A represents 8.

B represents $\left(\frac{10}{12}\right)$.

Step 2 is to rewrite the missing part of the given equation using the values for A and B.

$$8\left(\frac{10}{12}\right) = \left(\frac{10}{12}\right)8$$

The correct answer is $8\left(\frac{10}{12}\right) = \left(\frac{10}{12}\right)8$.

7.3   Associative and Distributive Properties

 Which of the following statements illustrates the Associative Property of Addition?

a) $(8x + 2) + 5y = 8x + (2 + 5y)$

b) $85 + 2 = 85y + 2y$

c) $\left(\dfrac{165}{5}\right) + 4x = \dfrac{(165 + 4x)}{5}$

d) $(5x - 2) + 20 = 20 + 5x - 2$

 The correct answer is statement a. Statement a correctly illustrates the Associative Property of Addition. This property states that grouping of terms is not relevant when performing addition. In statement a, $(8x + 2) + 5y = 8x + (2 + 5y)$, both sides of the equation will be equal, showing that grouping of terms is not relevant in addition. Statements b, c, and d do not illustrate the Associative Property of Addition.

 Using the Associative Property of Multiplication, fill in the missing part of the equation below.

$8(x \times 3y) = ?$

 The Associative Property of Multiplication states that grouping is not relevant in multiplication. The format for this property is shown below. Notice that the order is the same, but the parentheses have changed position.

$A(BC) = AB(C)$

Step 1 is to determine what terms represent A, B, and C in the given equation.

A represents 8.

B represents *x*.

C represents 3*y*.

Step 2 is to rewrite the missing part of the given equation using the values for *A*, *B*, and *C*.

$8(x \times 3y) = (8x)3y$

The correct answer is $8(x \times 3y) = (8x)3y$.

 Which of the following statements illustrates the Distributive Property?

a) $10(x - y) = 10x - 10y$

b) $3x + 11y = 3(x + 11y)$

c) $\dfrac{5}{8x(8y)}(3) = \dfrac{5x}{y} + 3$

d) $x(y)(z) = x + y + z$

 The correct answer is statement a. Statement a correctly illustrates the Distributive Property. This property states that multiplying (or distributing) terms in parentheses will not change the solution. In statement a, $10(x - y) = 10x - 10y$, both sides of the equation will be equal, showing that multiplying terms will not change the solution. Statements b, c, and d do not illustrate the Distributive Property.

 Use the Distributive Property to solve the following problem.

$6(x + y)$

The Distributive Property is shown in the format below.

$A(B + C) = AB + AC$

Step 1 is to determine what terms represent A, B, and C in the given equation.

A represents 6

B represents x

C represents y

Step 2 is to perform the multiplication operation.

$AB = 6(x) = 6x$

$AC = 6(y) = 6y$

Step 3 is to rewrite the problem.

$6(x + y) = 6x + 6y$

The answer is $6x + 6y$.

 Use the Distributive Property to solve the following problem.

$15(1 + 4)$

 The Distributive Property is shown in the format below.

$A(B + C) = AB + AC$

Step 1 is to determine what terms represent A, B, and C in the given equation.

A represents 15.

B represents 1.

C represents 4.

Step 2 is to perform the multiplication operation.

$AB = 15(1) = 15$

$AC = 15(4) = 60$

Step 3 is to rewrite the problem.

$15(1 + 4) = 15 + 60$

Step 4 is to simplify the answer.

$15 + 60 = 75$

The answer is 75. Notice that if the addition operation inside the parentheses was done first, the solution would have been the same. However, that would not be using the Distributive Property. This property will become more important when solving algebraic equations.

7.4   Simplifying Algebraic Expressions

 Simplify the following expression.

$10(5 + 4)$

 Step 1 is to use the Distributive Property to rewrite the problem.

$10(5 + 4) = 10(5) + 10(4)$

Step 2 is to perform the operation.

$10(5) + 10(4) = 50 + 40$

$50 + 40 = 90$

The correct answer is 90.

Simplify the following expression.

$$\frac{1}{2}x + \frac{3}{4}x + \frac{3}{2}x$$

 Step 1 is to group like terms. Since all the terms contain the variable x, all the terms in the expression are like terms. To make the problem easier, group the terms with the denominator "2" together.

$$\left(\frac{1}{2}x + \frac{3}{2}x\right) + \frac{3}{4}x$$

Step 2 is to perform the operation inside the parentheses.

$$\frac{1}{2}x + \frac{3}{2}x = \frac{4}{2}x = 2x$$

Step 3 is to rewrite the problem.

$$2x + \frac{3}{4}x$$

Step 4 is to perform the addition operation.

$$2x + \frac{3}{4}x = \frac{8}{4}x + \frac{3}{4}x = \frac{11}{4}x$$

The correct answer is $\frac{11}{4}x$.

 Simplify the following expression.

$$\frac{14y}{7} + 15 - 5y + 2$$

Step 1 is to group like terms. Group the terms containing the variable y together. Group the constants together.

$$\left(\frac{14y}{7} - 5y\right) + (15 + 2)$$

Step 2 is to perform the operation inside the parentheses.

$$\left(\frac{14y}{7} - 5y\right) = 2y - 5y$$

$$2y - 5y = -3y$$

Step 3 is to rewrite the problem.

$$-3y + (15 + 2)$$

Step 4 is to perform the operation inside the parentheses.

$$15 + 2 = 17$$

Step 5 is to rewrite the problem.

$$-3y + 17$$

Since the remaining terms are not like terms, the problem cannot be simplified any further.

The correct answer is $-3y + 17$.

Simplify the following expression.

$$25x + 25y + 5 + 25z - 56y - 8x + z - 1$$

Step 1 is to group like terms. Group the terms containing the same variable together. Group constants together.

$$(25x - 8x) + (25y - 56y) + (25z + z) + (5 - 1)$$

Step 2 is to perform the operation inside the parentheses for the variable x.

$$(25x - 8x) = 17x$$

Step 3 is to perform the operation inside the parentheses for the variable y.

$$(25y - 56y) = -31y$$

Step 4 is to perform the operation inside the parentheses for the variable z.

$$(25z + z) = 26z$$

Step 5 is to perform the operation inside the parentheses for the constants.

$$5 - 1 = 4$$

Step 6 is to rewrite the problem.

$$17x - 31y + 26z + 4$$

Since the remaining terms are not like terms, the problem cannot be simplified any further.

The correct answer is $17x - 31y + 26z + 4$.

Q Simplify the following expression.

$$3x^2 - 7 - 2x^2 + y$$

 Step 1 is to group like terms. Group the terms containing the variable x^2 together. Group the terms containing the variable y together. Group the constants together.

$$(3x^2 - 2x^2) + y - 7$$

Step 2 is to perform the operation inside the parentheses.

$3x^2 - 2x^2 = x^2$

Step 3 is to rewrite the problem.

$x^2 + y - 7$

Since the remaining terms are not like terms, the problem cannot be simplified any further.

The correct answer is $x^2 + y - 7$.

 Simplify the following expression.

$x(x + y) - 25$

 Step 1 is to use the Distributive Property to rewrite the problem.

$x(x) + x(y) - 25$

Step 2 is to perform the operation.

$x(x) + x(y) - 25 = x^2 + xy - 25$

Since the remaining terms are not like terms, the problem cannot be simplified any further.

The correct answer is $x^2 + xy - 25$.

7.5 Simplifying with Variables and Exponents

 Factor the following expression.

$10x - 10$

 Step 1 is to find the greatest common factor (GCF) for $10x$ and 10.

$10x = (1, 2, 5, 10)$

$10 = (1, 2, 5, 10)$

The GCF is 10.

Step 2 is to divide each term by the GCF.

$$\frac{10x}{10} = x$$

$$\frac{10}{10} = 1$$

Step 3 is to rewrite the problem. Place the GCF outside of the parentheses.

$10(x - 1)$

The correct answer is $10(x - 1)$.

 Factor the following expression.

$x^2 - x^3 + x^4$

Step 1 is to find the greatest common factor (GCF) for x^2, x^3, and x^4. When looking for the GCF for the coefficients, the coefficient for each term is 1, so there is no need to divide.

Step 2 is to find the term with the lowest exponent of the same variable, x.

x^2 has the lowest exponent.

Step 3 is to divide each term by x^2.

$$\frac{x^2}{x^2} = 1$$

$$-\frac{x^3}{x^2} = -x$$

$$\frac{x^4}{x^2} = x^2$$

Step 4 is to rewrite the problem. Place the lowest exponent outside of the parentheses.

$$x^2(x^2 - x + 1)$$

The correct answer is $x^2(x^2 - x + 1)$.

 Factor the following expression.

$2x^2 + 4x^4$

 Step 1 is to find the greatest common factor (GCF) for $2x^2$ and $4x^4$. When looking for the GCF for a variable, only use the coefficient.

$$2 = (1, 2)$$

$$4 = (1, 2, 4)$$

The GCF is 2.

Step 2 is to divide each term by the GCF.

$$\frac{2x^2}{2} = x^2$$

$$\frac{4x^4}{2} = 2x^4$$

Step 3 is to rewrite the problem. Place the GCF outside of the parentheses.

$$2(2x^4 + x^2)$$

Step 4 is to find the term with the lowest exponent of the same variable, x.

x^2 has the lowest exponent.

Step 5 is to divide each term by x^2.

$$\frac{x^2}{x^2} = 1$$

$$\frac{x^4}{x^2} = x^2$$

Step 6 is to rewrite the problem. Place the lowest exponent outside of the parentheses.

$$x^2(2)(x^2 + 1)$$

Step 7 is to multiply the GCF and the lowest exponent.

$$x^2(2) = 2x^2$$

The correct answer is $2x^2(2x^2 + 1)$.

Q Factor the following expression.

$6x^3y + 8x^2y - 4xy$

 A Step 1 is to find the greatest common factor (GCF) for $6x^3y$, $8x^2y$, and $4xy$. When looking for the GCF for a variable, only use the coefficient.

$$6 = (1, 2, 3, 6)$$

$$8 = (1, 2, 4, 8)$$

$4 = (1, 2, 4)$

The GCF is 2.

Step 2 is to divide each term by the GCF.

$$\frac{6x^3y}{2} = 3x^3y$$

$$\frac{8x^2y}{2} = 4x^2y$$

$$\frac{4xy}{2} = 2xy$$

Step 3 is to rewrite the problem. Place the GCF outside of the parentheses.

$$2(3x^3y + 4x^2y - 2xy)$$

Step 4 is to find the term with the lowest exponent of the same variable xy.

$2xy$ has the lowest exponent.

Step 5 is to divide each term by xy.

$$\frac{3x^3y}{xy} = 3x^2$$

$$\frac{4x^2y}{xy} = 4x$$

$$\frac{2xy}{xy} = 2$$

Step 6 is to rewrite the problem. Place the lowest exponent outside of the parentheses.

$$xy(2)(3x^2 + 4x - 2)$$

Step 7 is to multiply the GCF and the lowest exponent.

$$xy(2) = 2xy$$

The correct answer is $2xy(3x^2 + 4x - 2)$.

 Factor the following trinomial (also called a quadratic).

$$x^2 + 7x + 6$$

 Trinomials are in the format $ax^2 + bx + c$. Determine what terms represent a, b, and c in the given equation.

$$a = 1 \qquad b = 7 \qquad c = 6$$

Since the coefficient for x^2 is 1, use the following steps.

Step 1 is to list pairs of factors for c.

$$6 = (1, 6)\ (2, 3)$$

Step 2 is to find factors of 6 that equal b when added. The factors can be made negative if necessary.

$$6 + 1 = 7$$

Step 3 is to put the factors in the following format.

$$(x + 6)(x + 1)$$

Step 4 is to verify the answer by using the First–Outer–Inner–Last (FOIL) method.

$$(x + 6)(x + 1) = x(x) + x + 6x + 6$$

$$= x^2 + 7x + 6$$

The correct answer is $(x + 6)(x + 1)$.

 Factor the following polynomial.

$x^2 + 4x + 4$

 This polynomial is a perfect square trinomial. It can be identified using the following format.

$a^2 + 2(a)(b) + b^2$

If a and b are perfect squares and the middle term is a product of 2 times the square root of the first and last terms, then you can factor the polynomial using the perfect square trinomial method.

Step 1 is to identify a^2 and b^2 in the problem using the following format:

$a^2 + 2(a)(b) + b^2$

$a^2 = x^2$ $\qquad\qquad$ $b^2 = 4$

Step 2 is to take the square root of a^2 and b^2.

$\sqrt{x^2} = x$ $\qquad\qquad$ $\sqrt{4} = 2$

Step 3 is to put the results of Step 2 in the following format.

$(x + 2)(x + 2)$

Step 4 is to verify the solution by using the FOIL method.

$$(x + 2)(x + 2) = x^2 + 2x + 2x + 4$$

$$= x^2 + 4x + 4$$

The correct answer is $(x + 2)(x + 2)$, or $(x + 2)^2$.

Factor the following polynomial.

$27x^3 + 125$

 This polynomial is a sum of two cubes. It can be identified using the following format.

$a^3 + b^3$

If a and b are perfect cubes, then you can factor the polynomial using the sum of two cubes.

Step 1 is to identify a^3 and b^3 in the problem using the following format:

$a^3 + b^3$

$a^3 = 27x^3$ $\qquad\qquad$ $b^3 = 125$

Step 2 is to take the cube root of a^3 and b^3.

$\sqrt[3]{27x^3} = 3x$ $\qquad\qquad$ $\sqrt[3]{125} = 5$

Step 3 is to put the results of Step 2 into the following format.

$(3x + 5)([3x]^2 - (3x)(5) + [5]^2)$

Step 4 is to simplify the answer.

$(3x + 5)(9x^2 - 15x + 25)$

The correct answer is $(3x + 5)(9x^2 - 15x + 25)$.

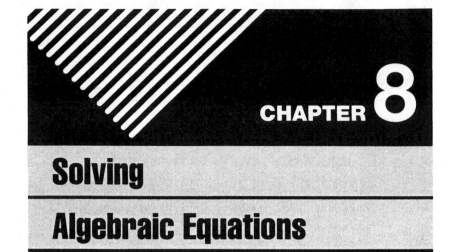

CHAPTER 8

Solving
Algebraic Equations

An **equation** is defined as a statement that two separate expressions are equal.

A **solution** to the equation is a number that makes the equation true when it is substituted for the variable. For example, in the equation $3x = 18$, 6 is the solution since $3(6) = 18$. Depending on the equation, there can be more than one solution. Equations with the same solutions are said to be **equivalent equations**. An equation without a solution is said to have a solution set that is the **empty** or **null** set and is represented by ϕ.

Replacing an expression within an equation by an equivalent expression will result in a new equation with solutions equivalent to the original equation. Given the equation below

$$3x + y + x + 2y = 15$$

by combining like terms, we get,

$$3x + y + x + 2y = 4x + 3y$$

Since these two expressions are equivalent, we can substitute the simpler form into the equation to get

$$4x + 3y = 15$$

Performing the same operation to both sides of an equation by the same expression will result in a new equation that is equivalent to the original equation.

8.1 Problem Solving with Addition Equations

 Find the solution(s) to the following problem.

$x + 17 = 12$

 Step 1 is to subtract 17 from both sides of the equation to isolate x.

$$x + 17 - 17 = 12 - 17$$

Step 2 is to simplify the equation.

$$x = 12 - 17$$

Step 3 is to solve the right side of the equation.

$$12 - 17 = -5$$

Step 4 is to rewrite the equation.

$$x = -5$$

Step 5 is to verify the solution by substituting for x.

$$(-5) + 17 = 12$$

The correct answer is $x = -5$.

 Find the solution(s) to the following problem.

$r^2 + 6 = 22$

Step 1 is to subtract 6 from both sides of the equation to isolate r^2.

$$r^2 + 6 - 6 = 22 - 6$$

Step 2 is to simplify the equation.

$$r^2 = 16$$

Step 3 is to take both the positive and negative square roots of both sides of the equation.

$$r = \pm\sqrt{16} = \pm 4$$

Step 4 is to verify the solutions by substituting for r.

$$(4)^2 + 6 = 22 \qquad (-4)^2 + 6 = 22$$

$$16 + 6 = 22 \qquad 16 + 6 = 22$$

The correct answer is $r = 4$ or $r = -4$.

Find the solution(s) to the following problem.

$$\frac{1}{2}m + \frac{2}{4}m + \frac{2}{16} = -\frac{1}{16}$$

Step 1 is to simplify the left side of the equation.

$$\frac{1}{2}m + \frac{2}{4}m + \frac{2}{16} = 1m + \frac{2}{16} = m + \frac{2}{16}$$

Step 2 is to rewrite the equation.

$$m + \frac{2}{16} = -\frac{1}{16}$$

Step 3 is to subtract $\dfrac{2}{16}$ from both sides of the equation to isolate m.

$$m + \frac{2}{16} - \frac{2}{16} = -\frac{1}{16} - \frac{2}{16}$$

Step 4 is to simplify the equation.

$$m = -\frac{1}{16} - \frac{2}{16}$$

Step 5 is to solve the right side of the equation.

$$-\frac{1}{16} - \frac{2}{16} = -\frac{3}{16}$$

Step 6 is to rewrite the equation.

$$m = -\frac{3}{16}$$

Step 7 is to verify the solution by substituting for m.

$$\frac{1}{2}\left(-\frac{3}{16}\right) + \frac{2}{4}\left(-\frac{3}{16}\right) + \frac{2}{16} = -\frac{1}{16}$$

$$-\frac{3}{32} - \frac{6}{64} + \frac{2}{16} = -\frac{1}{16}$$

$$-\frac{6}{64} - \frac{6}{64} + \frac{2}{16} = -\frac{1}{16}$$

$$-\frac{12}{64} + \frac{8}{64} = -\frac{1}{16}$$

$$-\frac{4}{64} = -\frac{1}{16}$$

$$-\frac{1}{16} = -\frac{1}{16}$$

The correct answer is $m = -\frac{3}{16}$.

 Find the solution(s) to the following problem.

$x^4 + 1 = 82$

 Step 1 is to subtract 1 from both sides of the equation to isolate x^4.

$x^4 + 1 - 1 = 82 - 1$

Step 2 is to simplify the equation.

$x^4 = 82 - 1$

Step 3 is to solve the right side of the equation.

$82 - 1 = 81$

Step 4 is to rewrite the equation.

$x^4 = 81$

Step 5 is to take both the positive and negative 4th roots of both sides of the equation.

$x = \pm\sqrt[4]{81} = \pm 3$

Step 6 is to rewrite the equation.

$x = 3$ or -3

Step 7 is to verify the solutions by substituting for x.

$(3)^4 + 1 = 82$ $(-3)^4 + 1 = 82$

The correct answer is $x = 3$ or $x = -3$.

 Find the solution(s) to the following problem.

$x^2 + 7x + 6 = 0$

 Step 1 is to factor the equation.

$x^2 + 7x + 6 = (x + 6)(x + 1)$

Step 2 is to write each term separately.

$(x + 6) = 0$

$(x + 1) = 0$

Step 3 is to solve each term.

$x + 6 = 0$ $x + 1 = 0$

$x = -6$ $x = -1$

Step 4 is to verify each solution in the original equation.

$(-6)^2 + 7(-6) + 6 = 0$ $(-1)^2 + 7(-1) + 6 = 0$

$36 + (-42) + 6 = 0$ $1 + (-7) + 6 = 0$

$0 = 0$ $0 = 0$

The correct solutions are $x = -6$ or $x = -1$.

 Find the solution(s) to the following problem.

$x^2 + 11x + 10 = 0$

 Step 1 is to factor the equation.

$x^2 + 11x + 10 = (x + 10)(x + 1)$

Step 2 is to write each term separately.

$(x + 10) = 0$

$(x + 1) = 0$

Step 3 is to solve each term.

$x + 10 = 0 \qquad x + 1 = 0$

$x = -10 \qquad\quad x = -1$

Step 4 is to verify each solution in the original equation.

$(-10)^2 + 11(-10) + 10 = 0 \qquad (-1)^2 + 11(-1) + 10 = 0$

$\qquad\quad 100 - 110 + 10 = 0 \qquad\qquad 1 - 11 + 10 = 0$

$\qquad\qquad\qquad\quad 0 = 0 \qquad\qquad\qquad\qquad 0 = 0$

The correct solutions are $x = -10$ or $x = -1$.

 Find the solution(s) to the following problem.

$x^2 + x - 12 = 0$

 Step 1 is to factor the equation.

$x^2 + x - 12 = (x + 4)(x - 3)$

Step 2 is to write each term separately.

$(x + 4) = 0$

$(x - 3) = 0$

Step 3 is to solve each term.

$x + 4 = 0 \qquad\qquad\qquad x - 3 = 0$

$\quad x = -4 \qquad\qquad\qquad\quad x = 3$

Step 4 is to verify each solution in the original equation.

$$(-4)^2 + (-4) - 12 = 0 \qquad (3)^2 + 3 - 12 = 0$$

$$16 - 4 - 12 = 0 \qquad 9 + 3 - 12 = 0$$

$$0 = 0 \qquad 0 = 0$$

The correct solutions are $x = -4$ or $x = 3$.

 Find the solution(s) to the following expression.

$2x^2 - 7x + 1$

 Step 1 is to factor the expression.

$2x^2 - 7x + 1$ cannot be factored.

Therefore, solutions to the expression $2x^2 - 7x + 1$ cannot be found by factoring.

 Find the solution(s) to the following problem.

$4x^2 + 16x + 16 = 0$

 Step 1 is to factor the equation.

$4x^2 + 16x + 16 = 4(x^2 + 4x + 4) = 4(x + 2)(x + 2)$

Step 2 is to write each term separately (we can drop the 4 because 4=0).

$(x + 2) = 0$

$(x + 2) = 0$

Step 3 is to solve each term; since they are the same, we only need to do it once.

$x + 2 = 0$

$x = -2$

Step 4 is to verify each solution. Since the solution for each of the terms above is the same, only one verification is needed.

$4(-2)^2 + 16(-2) + 16 = 0$

$4(4) - 32 + 16 = 0$

$0 = 0$

The correct solution is $x = -2$.

8.2 Problem Solving with Subtraction Equations

Find the solution(s) to the following problem.

$t^2 - 10 = 90$

Step 1 is to add 10 to both sides of the equation to isolate t^2.

$t^2 - 10 + 10 = 90 + 10$

Step 2 is to simplify the equation.

$t^2 = 90 + 10$

Step 3 is to solve the right side of the equation.

$90 + 10 = 100$

Step 4 is to rewrite the equation.

$t^2 = 100$

Step 5 is to take both the positive and negative square roots of both sides of the equation.

$t = \pm\sqrt{100} = \pm 10$

Step 6 is to rewrite the equation.

$t = 10$ or -10 (or ±10)

Step 7 is to verify the solutions by substituting for t.

$(10)^2 - 10 = 90$	$(-10)^2 - 10 = 90$
$100 - 10 = 90$	$100 - 10 = 90$
$90 = 90$	$90 = 90$

The correct answer is $t = 10$ or $t = -10$.

 Find the solution(s) to the following problem.

$2z - 3z - 1 = 23$

 Step 1 is to simplify the left side of the equation.

$2z - 3z - 1 = -z - 1$

Step 2 is to rewrite the equation.

$-z - 1 = 23$

Step 3 is to add 1 to both sides of the equation to isolate z.

$-z - 1 + 1 = 23 + 1$

Step 4 is to simplify the equation.

$-z = 23 + 1$

Step 5 is to solve the right side of the equation.

$23 + 1 = 24$

Step 6 is to rewrite the equation.

$-z = 24$

Step 7 is to divide both sides of the equation by -1.

$$\frac{-z}{-1} = z \qquad\qquad \frac{24}{-1} = -24$$

Step 8 is to rewrite the equation.

$z = -24$

Step 9 is to verify the solution by substituting for z.

$$2(-24) - 3(-24) - 1 = 23$$

$$-48 + 72 - 1 = 23$$

$$23 = 23$$

The correct answer is $z = -24$.

 Find the solution(s) to the following problem.

$$y - 1 = \sqrt{4} - 2$$

 Step 1 is to add 1 to both sides of the equation to isolate y.

$$y - 1 + 1 = \sqrt{4} - 2 + 1$$

Step 2 is to simplify the equation.

$$y = \sqrt{4} - 2 + 1$$

Step 3 is to solve the right side of the equation.

$$\sqrt{4} - 2 + 1 = 2 - 2 + 1 = 1$$

Step 4 is to rewrite the equation.

$$y = 1$$

Step 5 is to verify the solution by substituting for y.

$$1 - 1 = \sqrt{4} - 2$$

$$0 = 0$$

The correct answer is $y = 1$.

 Find the solution(s) to the following problem.

$$6y - x - 7 = 6y$$

 Step 1 is to add 7 to both sides of the equation to isolate x.

$$6y - x - 7 + 7 = 6y + 7$$

Step 2 is to simplify the equation.

$$6y - x = 6y + 7$$

Step 3 is to subtract $6y$ from both sides of the equation to isolate x.

$$6y - x - 6y = 6y + 7 - 6y$$

Step 4 is to simplify both sides of the equation.

$$-x = 7$$

Step 5 is to multiply both sides of the equation by -1.

$$-1(-x) = (-1)(7)$$

$$x = -7$$

Step 6 is to verify the solution by substituting for x.

$$6y - (-7) - 7 = 6y$$

$$-(-7) - 7 = 0$$

$$7 - 7 = 0$$

$$0 = 0$$

The correct answer is $x = -7$.

Find the solution(s) to the following problem.

$x^2 - 4x - 5 = 0$

Step 1 is to factor the equation.

$x^2 - 4x - 5 = (x - 5)(x + 1)$

Step 2 is to write each term separately.

$(x - 5) = 0$

$(x + 1) = 0$

Step 3 is to solve each term.

$x - 5 = 0 \qquad\qquad x + 1 = 0$

$x = 5 \qquad\qquad x = -1$

Step 4 is to verify each solution in the original equation.

$(5)^2 - 4(5) - 5 = 0 \qquad\qquad (-1)^2 - 4(-1) - 5 = 0$

$25 - (20) - 5 = 0 \qquad\qquad 1 + 4 - 5 = 0$

$0 = 0 \qquad\qquad 0 = 0$

The correct solutions are $x = 5$ or $x = -1$.

Find the solution(s) to the following problem.

$x^2 - 14x - 15 = 0$

Step 1 is to factor the equation.

$x^2 - 14x - 15 = (x - 15)(x + 1)$

Step 2 is to write each term separately.

$(x - 15) = 0$

$(x + 1) = 0$

Step 3 is to solve each term.

$x - 15 = 0$	$x + 1 = 0$
$x = 15$	$x = -1$

Step 4 is to verify each solution in the original equation.

$(15)^2 - 14(15) - 15 = 0$	$(-1)^2 - 14(-1) - 15 = 0$
$225 - 210 - 15 = 0$	$1 + 14 - 15 = 0$
$0 = 0$	$0 = 0$

The correct solutions are $x = 15$ or $x = -1$.

 Find the solution(s) to the following problem.

$5x^2 - 25x - 30 = 0$

 Step 1 is to factor the equation.

$5x^2 - 25x - 30 = 5(x^2 - 5x - 6) = 5(x - 6)(x + 1)$

Step 2 is to write each term separately (we can drop the 5 because 5=0).

$(x - 6) = 0$

$(x + 1) = 0$

Step 3 is to solve each term without the factored number.

$x - 6 = 0$	$x + 1 = 0$
$x = 6$	$x = -1$

Step 4 is to verify each solution in the original equation.

$$5(6)^2 - 25(6) - 30 = 0 \qquad 5(-1)^2 - 25(-1) - 30 = 0$$

$$180 - 150 - 30 = 0 \qquad 5 + 25 - 30 = 0$$

$$0 = 0 \qquad\qquad 0 = 0$$

The correct solutions are $x = 6$ or $x = -1$.

8.3 Problem Solving with Multiplication Equations

Find the solution(s) to the following problem.

$2x^2 + 5x + 2 = 0$

Step 1 is to factor the equation.

$(2x + 1)(x + 2) = 0$

Step 2 is to write each term separately.

$2x + 1 = 0$

$x + 2 = 0$

Step 3 is to solve each term.

$$2x + 1 = 0 \qquad\qquad x + 2 = 0$$

$$2x = -1 \qquad\qquad x = -2$$

$$x = -\frac{1}{2}$$

Step 4 is to verify the solutions in the original equation.

$$2\left(-\frac{1}{2}\right)^2 + 5\left(-\frac{1}{2}\right) + 2 = 0 \qquad 2(-2)^2 + 5(-2) + 2 = 0$$

$$2\left(\frac{1}{4}\right) - \frac{5}{2} + 2 = 0 \qquad\qquad 2(4) - 10 + 2 = 0$$

$$\frac{1}{2} - \frac{5}{2} + 2 = 0 \qquad\qquad 8 - 10 + 2 = 0$$

$$-\frac{4}{2} + 2 = 0 \qquad\qquad -2 + 2 = 0$$

$$0 = 0 \qquad\qquad 0 = 0$$

The correct solutions to the problem are $x = -\frac{1}{2}$ and $x = -2$.

 Find the solution(s) to the following problem.

$4x^2 - 2x - 2 = 0$

 Step 1 is to factor the equation.

$2(2x^2 - x - 1) = 2(x - 1)(2x + 1) = 0$

Step 2 is to write each term separately (we can drop the 2 because 2=0).

$x - 1 = 0$

$2x + 1 = 0$

Step 3 is to solve each term.

$$x - 1 = 0 \qquad\qquad 2x + 1 = 0$$

$$2x = -1$$

$$x = 1 \qquad\qquad x = -\frac{1}{2}$$

Step 4 is to verify the solutions.

$$4(1) - 2(1) - 2 = 0 \qquad 4\left(\frac{1}{4}\right) - 2\left(-\frac{1}{2}\right) - 2 = 0$$

$$4 - 2 - 2 = 0 \qquad\qquad 1 + 1 - 2 = 0$$

$$0 = 0 \qquad\qquad 0 = 0$$

The correct solutions to the problem are $x = 1$ or $x = -\frac{1}{2}$.

Find the solution(s) to the following problem.

$3x^2 - 4x + 1 = 0$

Step 1 is to factor the equation.

$(3x - 1)(x - 1) = 0$

Step 2 is to write each term separately.

$3x - 1 = 0$

$x - 1 = 0$

Step 3 is to solve each term.

$$3x - 1 = 0 \qquad\qquad x - 1 = 0$$

$$3x = 1 \qquad\qquad x = 1$$

$$x = \frac{1}{3}$$

Step 4 is to verify the solutions.

$$3\left(\frac{1}{3}\right)^2 - 4\left(\frac{1}{3}\right) + 1 = 0 \qquad\qquad 3(1)^2 - 4(1) + 1 = 0$$

$$3\left(\frac{1}{9}\right) - \frac{4}{3} + 1 = 0 \qquad\qquad -1 + 1 = 0$$

$$\frac{3}{9} - \frac{12}{9} + 1 = 0 \qquad\qquad 0 = 0$$

$$-\frac{9}{9} + 1 = 0$$

$$0 = 0$$

The correct solutions to the problem are $x = \dfrac{1}{3}$ or $x = 1$.

Find the solution(s) to the following problem.

$$\sqrt{4x} + 2 = 8$$

Step 1 is to isolate the variable $\sqrt{4x}$. To do this, subtract 2 from both sides of the equation.

$$\sqrt{4x} + 2 - 2 = 8 - 2$$

Step 2 is to simplify the left side of the equation.

$$\sqrt{4x} = 8 - 2$$

Step 3 is to solve the right side of the equation.

$$8 - 2 = 6$$

Step 4 is to rewrite the equation.

$$\sqrt{4x} = 6$$

Step 5 is to isolate the variable $4x$ by squaring both sides of the equation.

$$\sqrt{4x}\ \sqrt{4x} = 6(6)$$

$$4x = 36$$

Step 6 is to isolate the variable x by dividing both sides of the equation by 4.

$$\frac{4x}{4} = x \qquad\qquad \frac{36}{4} = 9$$

Step 7 is to rewrite the equation.

$$x = 9$$

Step 8 is to verify the equation by substituting for x.

$$\sqrt{4(9)} + 2 = 8$$

$$\sqrt{36} + 2 = 8$$

$$6 + 2 = 8$$

$$8 = 8$$

The correct solution is $x = 9$.

8.4   Problem Solving with Division Equations

Find the solution(s) to the following problem.

$$\frac{2}{x} = 10$$

Step 1 is to isolate the constants. To do this, multiply both sides of the equation by x.

$$\frac{2}{x}(x) = 2 \qquad\qquad 10(x) = 10x$$

Step 2 is to rewrite the equation.

$$2 = 10x$$

Step 3 is to isolate the variable x by dividing both sides of the equation by 10.

$$\frac{2}{10} = \frac{1}{5} \qquad\qquad \frac{10x}{10} = x$$

Step 4 is to rewrite the equation.

$$\frac{1}{5} = x$$

Step 5 is to verify the equation by substituting for x.

$$\frac{2}{\left(\frac{1}{5}\right)} = 10$$

$$2\left(\frac{5}{1}\right) = 10$$

$$10 = 10$$

The correct solution to this problem is $x = \frac{1}{5}$.

Find the solution(s) to the following problem.

$$\frac{2}{8} = \frac{1}{16}x$$

 Step 1 is to isolate the variable x by dividing both sides of the equation by $\frac{1}{16}$.

$$\frac{\left(\frac{2}{8}\right)}{\frac{1}{16}} = \frac{\left(\frac{1}{16}x\right)}{\frac{1}{16}}$$

$$\frac{2}{8} \times \frac{16}{1} = x$$

$$\frac{32}{8} = x$$

$$4 = x$$

Step 2 is to rewrite the equation.

$$4 = x$$

Step 3 is to verify the equation by substituting for x.

$$\frac{2}{8} = \frac{1}{16}(4)$$

$$\frac{2}{8} = \frac{4}{16}$$

$$\frac{2}{8} = \frac{2}{8}$$

The correct solution to this problem is $x = 4$.

 Find the solution(s) to the following problem.

$$\frac{1}{x+2} = 12$$

 Step 1 is to get rid of the fraction. To do this, multiply both sides of the equation by $x + 2$.

$$\frac{1}{x+2}(x+2) = 12(x+2)$$

Step 2 is to simplify the left side of the equation.

$$\frac{1}{x+2}(x+2) = 1$$

Step 3 is to rewrite the equation.

$$1 = 12(x+2)$$

Step 4 is to isolate the term $x + 2$ by dividing both sides of the equation by 12.

$$\frac{1}{12} = \frac{1}{12} \qquad\qquad \frac{12(x+2)}{12} = x+2$$

Step 5 is to rewrite the equation.

$$\frac{1}{12} = x + 2$$

Step 6 is to isolate the variable x. To do this, subtract 2 from both sides of the equation.

$$\frac{1}{12} - 2 = x + 2 - 2$$

Step 7 is to simplify the right side of the equation.

$$\frac{1}{12} - 2 = x$$

Step 8 is to simplify the left side of the equation.

$$\frac{1}{12} - 2 = -\frac{23}{12}$$

Step 9 is to rewrite the equation.

$$-\frac{23}{12} = x$$

Step 10 is to verify the solution by substituting for x.

$$\frac{1}{\left(-\frac{23}{12}\right) + 2} = 12$$

$$\frac{1}{\frac{1}{12}} = 12$$

$$12 = 12$$

The correct solution to this problem is $x = -\frac{23}{12}$.

Find the solution(s) to the following problem.

$$\frac{12}{(x-2)} = x - 2$$

Step 1 is to isolate the constants. To do this, multiply both sides of the equation by $(x - 2)$.

$$\frac{12}{x-2}(x-2) = (x-2)(x-2)$$

Step 2 is to simplify the left side of the equation.

$$\frac{12}{(x-2)}(x-2) = 12$$

Step 3 is to rewrite the equation.

$$12 = (x-2)(x-2)$$

Step 4 is to simplify the right side of the equation.

$$(x-2)(x-2) = x^2 - 4x + 4$$

Step 5 is to rewrite the equation.

$$12 = x^2 - 4x + 4$$

Step 6 is to subtract 12 from both sides of the equation.

$$0 = x^2 - 4x - 8$$

Step 7 is to factor the equation.

$x^2 - 4x - 8$ does not factor.

Therefore, solutions to the expression $x^2 - 4x - 8$ cannot be found by factoring.

Find the solution(s) to the following problem.

$$\frac{2}{x^2 + 3x - 2} = 1$$

Step 1 is to eliminate the fraction. To do this, multiply both sides of the equation by $x^2 + 3x - 2$.

$$\frac{2}{x^2 + 3x - 2}(x^2 + 3x - 2) = 1(x^2 + 3x - 2)$$

Step 2 is to simplify the left side of the equation.

$$\frac{2}{x^2 + 3x - 2}(x^2 + 3x - 2) = 2$$

Step 3 is to simplify the right side of the equation.

$$1(x^2 + 3x - 2) = x^2 + 3x - 2$$

Step 4 is to rewrite the equation.

$$2 = x^2 + 3x - 2$$

Step 5 is to subtract 2 from both sides of the equation.

$$2 - 2 = x^2 + 3x - 2 - 2$$

Step 6 is to simplify the equation.

$$0 = x^2 + 3x - 4$$

Step 7 is to factor the equation.

$$0 = (x + 4)(x - 1)$$

Step 8 is to write each term separately.

$$x + 4 = 0$$

$$x - 1 = 0$$

Step 9 is to solve each term.

$$x + 4 = 0 \qquad\qquad x - 1 = 0$$

$$x = -4 \qquad\qquad x = 1$$

Step 10 is to verify each solution by substituting for x.

solution 1
$$\frac{2}{x^2 + 3x - 2} = 1$$

$$\frac{2}{(-4)^2 + 3(-4) - 2} = 1$$

$$\frac{2}{16 - 12 - 2} = 1$$

$$1 = 1$$

solution 2
$$\frac{2}{x^2 + 3x - 2} = 1$$

$$\frac{2}{(1)^2 + 3(1) - 2} = 1$$

$$\frac{2}{1 + 3 - 2} = 1$$

$$1 = 1$$

The correct solutions to the problem are $x = -4$ or $x = 1$.

Find the solution(s) to the following problem.

$$\frac{1}{x^2 - \frac{1}{2}x - 1} = 2$$

 Step 1 is to eliminate the fraction. To do this, multiply both sides of the equation by $\left(x^2 - \frac{1}{2}x - 1 \right)$.

$$\frac{1}{x^2 - \frac{1}{2}x - 1}\left(x^2 - \frac{1}{2}x - 1 \right) = 2\left(x^2 - \frac{1}{2}x - 1 \right)$$

Step 2 is to simplify the left side of the equation.

$$\frac{1}{x^2 - \frac{1}{2}x - 1}\left(x^2 - \frac{1}{2}x - 1 \right) = 1$$

Step 3 is to simplify the right side of the equation.

$$2\left(x^2 - \frac{1}{2}x - 1 \right) = 2x^2 - x - 2$$

Step 4 is to rewrite the equation.

$$1 = 2x^2 - x - 2$$

Step 5 is to subtract 1 from both sides of the equation.

$$1 - 1 = 2x^2 - x - 2 - 1$$

Step 6 is to simplify the equation.

$$0 = 2x^2 - x - 3$$

Step 7 is to factor the equation.

$$0 = (2x - 3)(x + 1)$$

Step 8 is to write each term separately.

$$2x - 3 = 0$$

$$x + 1 = 0$$

Step 9 is to solve each term.

$$2x - 3 = 0 \qquad\qquad x + 1 = 0$$

$$x = \frac{3}{2} \qquad\qquad x = -1$$

Step 10 is to verify each solution by substituting for x.

solution 1
$$\frac{1}{x^2 - \frac{1}{2}x - 1} = 2$$

$$\frac{1}{\left(\frac{3}{2}\right)^2 - \frac{1}{2}\left(\frac{3}{2}\right) - 1} = 2$$

$$\frac{1}{\frac{9}{4} - \frac{3}{4} - 1} = 2$$

$$\frac{1}{\frac{1}{2}} = 2$$

$$2 = 2$$

solution 2

$$\frac{1}{x^2 - \frac{1}{2}x - 1} = 2$$

$$\frac{1}{(-1)^2 - \frac{1}{2}(-1) - 1} = 2$$

$$\frac{1}{1 + \frac{1}{2} - 1} = 2$$

$$\frac{1}{\frac{1}{2}} = 2$$

$$2 = 2$$

The correct solutions to the problem are $x = \frac{3}{2}$ or $x = -1$.

8.5   Problem Solving with Mixed Equations

 Find the solution(s) to the following problem.

$2(r^2 - 5) = 22$

 Step 1 is to divide both sides of the equation by 2.

$$\frac{\left[2\left(r^2 - 5\right)\right]}{2} = \frac{22}{2}$$

$$r^2 - 5 = 11$$

Step 2 is to add 5 to both sides of the equation to isolate r^2.

$r^2 - 5 + 5 = 11 + 5$

Step 3 is to simplify the left side of the equation.

$r^2 = 11 + 5$

Step 4 is to simplify the right side of the equation.

$11 + 5 = 16$

Step 5 is to rewrite the equation.

$r^2 = 16$

Step 6 is to take both the positive and negative square roots of both sides of the equation.

$r = \pm \sqrt{16} = \pm 4$

Step 7 is to rewrite the equation.

$r = 4, -4$

Step 8 is to verify the solutions by substituting for r.

$2[(4)^2 - 5)] = 22$ $\qquad$ $2[(-4)^2 - 5)] = 22$

$2[16 - 5] = 22$ $\qquad$ $2[16 - 5] = 22$

$2[11] = 22$ $\qquad$ $2[11] = 22$

$22 = 22$ $\qquad$ $22 = 22$

The correct answers are $r = 4$ or $r = -4$.

 Find the solution(s) to the following problem.

$x^2 + 5x + 6 = 0$

Step 1 is to factor the equation.

$x^2 + 5x + 6 = (x + 3)(x + 2)$

Step 2 is to write each term separately.

$(x + 3) = 0$

$(x + 2) = 0$

Step 3 is to solve each term.

$x + 3 = 0$	$x + 2 = 0$
$x = -3$	$x = -2$

Step 4 is to verify each solution.

$(-3)^2 + 5(-3) + 6 = 0$	$(-2)^2 + 5(-2) + 6 = 0$
$9 + (-15) + 6 = 0$	$4 + (-10) + 6 = 0$
$0 = 0$	$0 = 0$

The correct solutions are $x = -3$ or $x = -2$.

Find the solution(s) to the following problem.

$x^2 + 15x - 16 = 0$

Step 1 is to factor the equation.

$x^2 + 15x - 16 = (x - 1)(x + 16)$

Step 2 is to write each term separately.

$(x - 1) = 0$

$(x + 16) = 0$

Step 3 is to solve each term.

$$x - 1 = 0 \qquad\qquad x + 16 = 0$$

$$x = 1 \qquad\qquad x = -16$$

Step 4 is to verify each solution.

$$(1)^2 + 15(1) - 16 = 0 \quad (-16)^2 + 15(-16) - 16 = 0$$

$$1 + 15 - 16 = 0 \qquad\qquad 256 - 240 - 16 = 0$$

$$0 = 0 \qquad\qquad\qquad 0 = 0$$

The correct solutions are $x = 1$ or $x = -16$.

Find the solution(s) to the following problem.

$$\frac{8}{5}x + \frac{1}{5}x - \frac{1}{2} = \frac{1}{10}$$

Step 1 is to simplify the left side of the equation.

$$\frac{8}{5}x + \frac{1}{5}x - \frac{1}{2} = \frac{9}{5}x - \frac{1}{2}$$

Step 2 is to rewrite the equation.

$$\frac{9}{5}x - \frac{1}{2} = \frac{1}{10}$$

Step 3 is to add $\frac{1}{2}$ to both sides of the equation to isolate $\frac{9}{5}x$.

$$\frac{9}{5}x - \frac{1}{2} + \frac{1}{2} = \frac{1}{10} + \frac{1}{2}$$

Step 4 is to simplify the left side of the equation.

$$\frac{9}{5}x = \frac{1}{10} + \frac{1}{2}$$

Step 5 is to simplify the right side of the equation.

$$\frac{1}{10} + \frac{1}{2} = \frac{6}{10}$$

Step 6 is to rewrite the equation.

$$\frac{9}{5}x = \frac{6}{10}$$

Step 7 is to multiply both sides of the equation by $\frac{5}{9}$ to isolate x.

$$\frac{9}{5}x\left(\frac{5}{9}\right) = x \qquad\qquad \frac{6}{10}\left(\frac{5}{9}\right) = \frac{1}{3}$$

Step 8 is to rewrite the equation.

$$x = \frac{1}{3}$$

Step 9 is to verify the solution by substituting for x.

$$\frac{8}{5}\left(\frac{1}{3}\right) + \frac{1}{5}\left(\frac{1}{3}\right) - \frac{1}{2} = \frac{1}{10}$$

$$\frac{8}{15} + \frac{1}{15} - \frac{1}{2} = \frac{1}{10}$$

$$\frac{9}{15} - \frac{1}{2} = \frac{1}{10}$$

$$\frac{18}{30} - \frac{15}{30} = \frac{1}{10}$$

$$\frac{3}{30} = \frac{1}{10}$$

$$\frac{1}{10} = \frac{1}{10}$$

The correct answer is $x = \dfrac{1}{3}$.

 Find the solution(s) to the following problem.

$5x - 3x = \sqrt{144}$

 Step 1 is to simplify the left side of the equation.

$5x - 3x = 2x$

Step 2 is to rewrite the equation.

$2x = \sqrt{144}$

Step 3 is to isolate the variable x by dividing both sides of the equation by 2.

$$\frac{2x}{2} = x \qquad\qquad \frac{\sqrt{144}}{2} = \frac{12}{2}$$

Step 4 is to simplify the right side of the equation.

$$\frac{12}{2} = 6$$

Step 5 is to rewrite the equation.

$x = 6$

Step 6 is to verify the solution by substituting for x.

$$5x - 3x = \sqrt{144}$$

$$5(6) - 3(6) = \sqrt{144}$$

$$30 - 18 = 12$$

$$12 = 12$$

The correct solution to the equation is $x = 6$.

 Find the solution(s) to the following problem.

$$4y^2 + 4y = 0$$

 Step 1 is to factor the equation.

$$4y^2 + 4y = 0$$

$$4y(y + 1) = 0$$

Step 2 is to write each term separately.

$$y + 1 = 0$$

$$4y = 0$$

Step 3 is to solve each term.

$y + 1 = 0$	$4y = 0$
$y = -1$	$y = 0$

Step 4 is to verify the solutions.

$4(-1)^2 + 4(-1) = 0$	$4(0)^2 + 4(0) = 0$
$4(1) + 4(-1) = 0$	$0 + 0 = 0$
$4 - 4 = 0$	$0 = 0$
$0 = 0$	

The correct solutions to the problem are $y = -1$ or $y = 0$.

 Find the solution(s) to the following problem.

$2x^2 - 3x - 2 = 0$

 Step 1 is to factor the equation.

$(2x + 1)(x - 2) = 0$

Step 2 is to write each term separately.

$2x + 1 = 0$

$x - 2 = 0$

Step 3 is to solve each term.

$2x + 1 = 0$	$x - 2 = 0$
$2x = -1$	$x = 2$
$x = -\dfrac{1}{2}$	

Step 4 is to verify the solutions.

$$2\left(-\frac{1}{2}\right)^2 - 3\left(-\frac{1}{2}\right) - 2 = 0 \qquad 2(2)^2 - 3(2) - 2 = 0$$

$$2\left(\frac{1}{4}\right) + \frac{3}{2} - 2 = 0 \qquad 2(4) - 6 - 2 = 0$$

$$\frac{1}{2} + \frac{3}{2} - 2 = 0 \qquad 8 - 6 - 2 = 0$$

$$\frac{4}{2} - 2 = 0 \qquad 8 - 8 = 0$$

$$0 = 0 \qquad 0 = 0$$

The correct solutions to the problem are $x = -\dfrac{1}{2}$ or $x = 2$.

Find the solution(s) to the following problem.

$$\frac{11}{x-4} = 3$$

Step 1 is eliminate the fraction. To do this, multiply both sides of the equation by $x - 4$.

$$\frac{11}{x-4}(x-4) = 3(x-4)$$

Step 2 is to simplify the left side of the equation.

$$\frac{11}{x-4}(x-4) = 11$$

Step 3 is to rewrite the equation.

$$11 = 3(x-4)$$

Step 4 is to isolate the term $x - 4$ by dividing both sides of the equation by 3.

$$\frac{11}{3} = \frac{11}{3} \qquad\qquad \frac{3(x-4)}{3} = x - 4$$

Step 5 is to rewrite the equation.

$$\frac{11}{3} = x - 4$$

Step 6 is to isolate the variable x. To do this, add 4 to both sides of the equation.

$$\frac{11}{3} + 4 = x - 4 + 4$$

Step 7 is to simplify the right side of the equation.

$$\frac{11}{3} + 4 = x$$

Step 8 is to simplify the left side of the equation.

$$\frac{11}{3} + 4 = \frac{11}{3} + \frac{12}{3} = \frac{23}{3}$$

Step 9 is to rewrite the equation.

$$\frac{23}{3} = x$$

Step 10 is to verify the solution by substituting for x.

$$\frac{11}{\frac{23}{3} - 4} = 3$$

$$\frac{11}{\frac{11}{3}} = 3$$

$$3 = 3$$

The correct solution to this problem is $x = \dfrac{23}{3}$.

Find the solution(s) to the following problem.

$$\frac{8}{(x-5)} = x + 2$$

 Step 1 is to isolate the constants. To do this, multiply both sides of the equation by $(x - 5)$.

$$\frac{8}{(x-5)}(x - 5) = (x + 2)(x - 5)$$

Step 2 is to simplify the left side of the equation.

$$\frac{8}{(x-5)}(x - 5) = 8$$

Step 3 is to rewrite the equation.

$$8 = (x + 2)(x - 5)$$

Step 4 is to simplify the right side of the equation.

$$(x + 2)(x - 5) = x^2 - 3x - 10$$

Step 5 is to rewrite the equation.

$$8 = x^2 - 3x - 10$$

Step 6 is to subtract 8 from both sides of the equation.

$$0 = x^2 - 3x - 18$$

Step 7 is to factor the equation.

$$x^2 - 3x - 18 = (x - 6)(x + 3)$$

Step 8 is to write each term separately.

$$x - 6 = 0$$

$x + 3 = 0$

Step 9 is to solve each term.

$$x - 6 = 0 \qquad\qquad x + 3 = 0$$

$$x = 6 \qquad\qquad x = -3$$

Step 10 is to verify each solution by substituting for x.

$$\frac{8}{(6-5)} = 6 + 2 \qquad\qquad \frac{8}{(-3-5)} = -3 + 2$$

$$\frac{8}{1} = 8 \qquad\qquad \frac{8}{-8} = -1$$

$$8 = 8 \qquad\qquad -1 = -1$$

The solutions to the problem are $x = 6$ or $x = -3$.

CHAPTER 9

Inequalities

An inequality is a statement where the value of one quantity or expression is greater than (>), less than (<), greater than or equal to (≥), less than or equal to (≤), or not equal to (≠) that of another. E.g.,

$$5 > 4$$

The expression above means that the value of 5 is greater than the value of 4.

A **conditional inequality** is an inequality whose validity depends on the values of the variables in the sentence. That is, certain values of the variables will make the sentence true, and others will make it false. $3 - y > 3 + y$ is a conditional inequality for the set of real numbers, since it is true for any replacement less than zero and false for all others.

$x + 5 > x + 2$ is an **absolute inequality** for the set of real numbers, meaning that for any real value x, the expression on the left is greater than the expression on the right.

$5y < 2y + y$ is inconsistent for the set of non-negative real numbers. For any y greater than 0 the sentence is always false. A sentence is inconsistent if it is always false when its variables assume allowable values.

The solution of a given inequality in one variable x consists of all values of x for which the inequality is true.

The graph of an inequality in one variable is represented by either a ray or a line segment on the real number line.

The endpoint is not a solution if the variable is strictly less than or greater than a particular value. E.g.,

$x > 2$

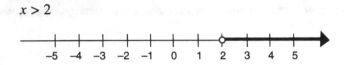

2 is not a solution and should be represented as shown.

The endpoint is a solution if the variable is either (1) less than or equal to or (2) greater than or equal to a particular value. E.g.,

$5 > x \geq 2$

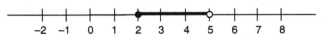

In this case 2 is the solution and should be represented as shown.

9.1 Properties of Inequalities

If x and y are real numbers then one and only one of the following statements is true.

$x > y$, $x = y$, or $x < y$.

This is the order property of real numbers.

If a, b and c are real numbers:

A) If $a < b$ and $b < c$, then $a < c$.

B) If $a > b$ and $b > c$, then $a > c$.

This is the transitive property of inequalities.

If *a*, *b*, and *c* are real numbers and $a > b$, then $a + c > b + c$ and $a - c > b - c$. This is the **addition property of inequality**.

Two inequalities are said to have the same **sense** if their signs of inequality point in the same direction.

The sense of an inequality remains the same if both sides are multiplied or divided by the same positive real number.

9.2 Inequalities with Addition and Subtraction

Find the solution to the following inequality.

$x + 35 > 100$

Step 1 is to isolate the variable by subtracting 35 from both sides of the inequality.

$x + 35 - 35 > 100 - 35$

Step 2 is to simplify the left side of the inequality.

$x > 100 - 35$

Step 3 is to simplify the right side of the inequality.

$100 - 35 = 65$

Step 4 is to rewrite the inequality.

$x > 65$

Step 5 is to verify the solution. To do this, choose any number that is greater than 65 and substitute that for the variable *x*.

$$70 + 35 > 100$$

$$105 > 100$$

The solution to the inequality is $x > 65$.

 Find the solution to the following inequality.

$$y - 12 > 3$$

 Step 1 is to isolate the variable by adding 12 to both sides of the inequality.

$$y - 12 + 12 > 3 + 12$$

Step 2 is to simplify the left side of the inequality.

$$y > 3 + 12$$

Step 3 is to simplify the right side of the inequality.

$$3 + 12 = 15$$

Step 4 is to rewrite the inequality.

$$y > 15$$

Step 5 is to verify the solution. To do this, choose any number that is greater than 15 and substitute that for the variable y.

$$22 - 12 > 3$$

$$10 > 3$$

The solution to the inequality is $y > 15$.

 Find the solution to the following inequality.

$$\frac{1}{2} + z < 4$$

Step 1 is to isolate the variable by subtracting $\dfrac{1}{2}$ from both sides of the inequality.

$$\frac{1}{2} + z - \frac{1}{2} < 4 - \frac{1}{2}$$

Step 2 is to simplify the left side of the inequality.

$$z < 4 - \frac{1}{2}$$

Step 3 is to simplify the right side of the inequality.

$$4 - \frac{1}{2} = \frac{7}{2}$$

Step 4 is to rewrite the inequality.

$$z < \frac{7}{2}$$

Step 5 is to verify the solution. To do this, choose any number that is less than $\dfrac{7}{2}$ and substitute that for the variable z.

$$\frac{1}{2} + \frac{2}{2} < 4$$

$$\frac{3}{2} < 4$$

The solution to the inequality is $z < \dfrac{7}{2}$.

Find the solution to the following inequality.

$$\frac{1}{4} + x \geq 1$$

 Step 1 is to isolate the variable by subtracting $\frac{1}{4}$ from both sides of the inequality.

$$\frac{1}{4} + x - \frac{1}{4} \geq 1 - \frac{1}{4}$$

Step 2 is to simplify the left side of the inequality.

$$x \geq 1 - \frac{1}{4}$$

Step 3 is to simplify the right side of the inequality.

$$1 - \frac{1}{4} = \frac{3}{4}$$

Step 4 is to rewrite the inequality.

$$x \geq \frac{3}{4}$$

Step 5 is to verify the solution. To do this, choose any number that is greater than or equal to $\frac{3}{4}$ and substitute that for the variable x.

$$\frac{1}{4} + \frac{3}{4} \geq 1$$

$$1 \geq 1$$

The solution to the inequality is $x \geq \dfrac{3}{4}$.

 Find the solution to the following inequality.

$y - 45 \geq 10$

 Step 1 is to isolate the variable by adding 45 to both sides of the inequality.

$y - 45 + 45 \geq 10 + 45$

Step 2 is to simplify the left side of the inequality.

$y \geq 10 + 45$

Step 3 is to simplify the right side of the inequality.

$10 + 45 = 55$

Step 4 is to rewrite the inequality.

$y \geq 55$

Step 5 is to verify the solution. To do this, choose any number that is greater than or equal to 55 and substitute that for the variable y.

$56 - 45 \geq 10$

$11 \geq 10$

The solution to the inequality is $y \geq 55$.

 Find the solution to the following inequality.

$5x + 7x + 3 \leq 25$

 Step 1 is to isolate the variable by subtracting 3 from both sides of the inequality.

$$5x + 7x + 3 - 3 \le 25 - 3$$

Step 2 is to simplify the left side of the inequality.

$$5x + 7x = 12x$$

Step 3 is to rewrite the inequality.

$$12x \le 25 - 3$$

Step 4 is to simplify the right side of the inequality.

$$25 - 3 = 22$$

Step 5 is to rewrite the inequality.

$$12x \le 22$$

Step 6 is to divide both sides of the inequality by 12.

$$\frac{12x}{12} = x$$

$$\frac{22}{12} = \frac{11}{6}$$

Step 7 is to rewrite the inequality.

$$x \le \frac{11}{6}$$

Step 8 is to verify the solution. To do this, choose any number that is less than or equal to $\frac{11}{6}$ and substitute that for the variable x.

$$5(1) + 7(1) + 3 \le 25$$

$$5 + 7 + 3 \le 25$$

$$15 \leq 25$$

The solution to the inequality is $x \leq \dfrac{11}{6}$.

9.3   Inequalities with Multiplication and Division

Find the solution to the following problem.

$$\frac{y}{10} < 7$$

Step 1 is to isolate the variable. To do this, multiply both sides of the inequality by 10.

$$\left(\frac{y}{10}\right)(10) < (7)(10)$$

Step 2 is to simplify the left side of the inequality.

$$\left(\frac{y}{10}\right)(10) = y$$

Step 3 is to simplify the right side of the inequality.

$$7(10) = 70$$

Step 4 is to rewrite the inequality.

$$y < 70$$

Step 5 is to verify the solution. To do this, choose any number that is less than 70 and substitute that for the variable y.

$$\frac{1}{10}(50) < 7$$

$$5 < 7$$

The correct answer to the inequality is $y < 70$.

 Find the solution to the following problem.

$$10x + 20x - 20 > 30$$

 Step 1 is to simplify the left side of the inequality.

$$10x + 20x - 20 = 30x - 20$$

Step 2 is to rewrite the inequality.

$$30x - 20 > 30$$

Step 3 is to isolate the variable. To do this, add 20 to both sides of the inequality.

$$30x - 20 + 20 > 30 + 20$$

Step 4 is to simplify the left side of the inequality.

$$30x > 30 + 20$$

Step 5 is to simplify the right side of the inequality.

$$30 + 20 = 50$$

Step 6 is to rewrite the inequality.

$$30x > 50$$

Step 7 is to divide both sides of the inequality by 30 to isolate the variable.

$$\frac{30x}{30} = x \qquad \frac{50}{30} = \frac{5}{3}$$

Step 8 is to rewrite the inequality.

$$x > \frac{5}{3}$$

Step 9 is to verify the solution. To do this, choose any number that is greater than $\frac{5}{3}$ and substitute that for the variable x.

$$10(5) + 20(5) - 20 > 30$$

$$50 + 100 - 20 > 30$$

$$130 > 30$$

The correct answer to the inequality is $x > \frac{5}{3}$.

 Find the solution to the following problem.

$7x - 6 \leq 15$

 Step 1 is to isolate the variable. To do this, add 6 to both sides of the inequality.

$7x - 6 + 6 \leq 15 + 6$

Step 2 is to simplify the left side of the inequality.

$7x \leq 15 + 6$

Step 3 is to simplify the right side of the inequality.

$15 + 6 = 21$

Step 4 is to rewrite the inequality.

$7x \leq 21$

Step 5 is to divide both sides of the inequality by 7 to isolate the variable.

$$\frac{7x}{7} = x \qquad\qquad \frac{21}{7} = 3$$

Step 6 is to rewrite the inequality.

$$x \le 3$$

Step 7 is to verify the solution. To do this, choose any number that is less than or equal to 3 and substitute that for the variable x.

$$7(3) - 6 \le 15$$

$$21 - 6 \le 15$$

$$15 \le 15$$

The correct answer to the inequality is $x \le 3$.

 Find the solution to the following problem.

$$\frac{2}{3}y \le 9$$

 Step 1 is to isolate the variable. To do this, multiply both sides of the inequality by $\frac{3}{2}$.

$$\frac{2}{3}y\left(\frac{3}{2}\right) \le 9\left(\frac{3}{2}\right)$$

Step 2 is to simplify the left side of the inequality.

$$y \le 9\left(\frac{3}{2}\right)$$

Step 3 is to simplify the right side of the inequality.

$$9\left(\frac{3}{2}\right) = \frac{27}{2}$$

Step 4 is to rewrite the inequality.

$$y \le \frac{27}{2}$$

Step 5 is to verify the solution. To do this, choose any number that is less than or equal to $\frac{27}{2}$ and substitute that for the variable y.

$$\frac{2}{3}(1) \le 9$$

$$\frac{2}{3} \le 9$$

The correct answer to the inequality is $x \le \frac{27}{2}$.

 Find the solution to the following problem.

$45y - 20 + 15y \ge 40$

 Step 1 is to simplify the left side of the inequality.

$45y - 20 + 15y = 60y - 20$

Step 2 is to rewrite the inequality.

$60y - 20 \ge 40$

Step 3 is to isolate the variable. To do this, add 20 to both sides of the inequality.

$$60y - 20 + 20 \geq 40 + 20$$

Step 4 is to simplify the left side of the inequality.

$$60y \geq 40 + 20$$

Step 5 is to simplify the right side of the inequality.

$$40 + 20 = 60$$

Step 6 is to rewrite the inequality.

$$60y \geq 60$$

Step 7 is to divide both sides of the inequality by 60 to isolate the variable y.

$$\frac{60y}{60} = y \qquad\qquad \frac{60}{60} = 1$$

Step 8 is to rewrite the inequality.

$$y \geq 1$$

Step 9 is to verify the solution. To do this, choose any number that is greater than or equal to 1 and substitute that for the variable y.

$$45(2) - 20 + 15(2) \geq 40$$

$$90 - 20 + 30 \geq 40$$

$$100 \geq 40$$

The correct answer to the inequality is $y \geq 1$.

 Find the solution to the following problem.

$$-12y \leq -72$$

Step 1 is to isolate the variable. To do this, divide both sides of the inequality by 12.

$$\frac{(-12y)}{12} \le \frac{-72}{12}$$

Step 2 is to simplify the left side of the inequality.

$$\frac{(-12y)}{12} = -y$$

Step 3 is to simplify the right side of the inequality.

$$\frac{-72}{12} = -6$$

Step 4 is to rewrite the inequality.

$$-y \le -6$$

Step 5 is to divide both sides of the inequality by −1.

$$\frac{-y}{-1} = y \qquad\qquad \frac{-6}{-1} = 6$$

Step 6 is to rewrite the inequality.

$$y \le 6$$

In Step 7, since we divided by a negative number, the ≤ sign must be reversed.

$$y \ge 6$$

Step 8 is to verify the solution. To do this, choose any number that is greater than or equal to 6 and substitute that for the variable *y*.

$-12(10) \leq -72$

$-120 \leq -72$

The correct answer to the inequality is $y \geq 6$.

CHAPTER 10

Word Problems

One of the main problems students have in mathematics involves solving word problems. The secret to solving these problems is being able to convert words into numbers and variables in the form of an algebraic equation.

The easiest way to approach a word problem is to read the question and ask yourself what you are trying to find. This unknown quantity can be represented by a variable.

Next, determine how the variable relates to the other quantities in the problem. More than likely, these quantities can be explained in terms of the original variable. If not, a separate variable may have to be used to represent a quantity.

Using these variables and the relationships determined among them, an equation can be written. Solve for a particular variable and then plug this number in for each relationship that involves this variable in order to find any unknown quantities.

Lastly, reread the problem to be sure that you have answered the questions correctly and fully.

10.1 Algebraic

The following illustrates how to formulate an equation and solve the problem.

Find two consecutive odd integers whose sum is 36.

Let x = the first odd integer

Let $x + 2$ = the second odd integer

The sum of the two numbers is 36. Therefore,

$$x + (x + 2) = 36$$

Simplifying,

$$2x + 2 = 36$$

$$2x = 34$$

$$x = 17$$

Substituting 17 for x, we find the second odd integer $= (x + 2) = (17 + 2) = 19$. Therefore, we find that the two consecutive odd integers whose sum is 36 are 17 and 19, respectively.

10.2 Writing Algebraic Equations

Rewrite the following phrase as an algebraic expression. Do not solve.

Twelve is subtracted from a number. It is then divided by 4. The quotient equals 5. What is the number?

Step 1 is to represent the unknown quantity. In this problem, the unknown quantity is the missing number.

Let x = the missing number

Step 2 is to write the left side of the equation. It is known that if 12 is subtracted from the missing number, then divided by 4, it will equal the given quotient. So, the left side of the equation will be:

$$\frac{(x-12)}{4}$$

Step 3 is to write the right side of the equation. It is known that the quotient is 5. Therefore, the right side of the equation will be:

5

Step 4 is to write both sides of the equation together.

$$\frac{(x-12)}{4}=5$$

The correct answer is $\frac{(x-12)}{4}=5$.

 Rewrite the following phrase as an algebraic expression. Do not solve.

Jimmy received $5 from his Uncle Jake for raking the yard. Combined with his savings, Jimmy has $80 toward buying a new bicycle. How much does Jimmy have in savings?

A Step 1 is to represent the unknown quantity. In this problem, the unknown quantity is the amount of money that Jimmy has in savings.

Let x = the amount of money in Jimmy's savings

Step 2 is to write the left side of the equation. It is known that the amount Jimmy received for raking the yard plus the amount he has in savings will equal the total amount. Therefore, the left side of the equation will be:

$5 + x

Step 3 is to write the right side of the equation. It is known that the total amount is $80. Therefore, the right side of the equation will be:

$80

Step 4 is to write both sides of the equation together.

$5 + x = $80

The correct answer is $5 + x = $80.

 Rewrite the following phrase as an algebraic expression. Do not solve.

The High Point Company makes plastic toy compasses. This year's profit will be twice as much as last year's profit plus $1,200. If this year's profit is $100,000, how much was last year's profit?

 Step 1 is to represent the unknown quantity. In this problem, the unknown quantity is last year's profit.

Let x = last year's profit

Step 2 is to write the left side of the equation. It is known that last year's profit times 2 plus $1,200 will equal this year's profit. Therefore, the left side of the equation will be:

$2x + $1,200

Step 3 is to write the right side of the equation. It is known that this year's profit is $100,000. Therefore, the right side of the equation will be:

$100,000

Step 4 is to write both sides of the equation together.

$2x + $1,200 = $100,000

The correct answer is $2x + \$1,200 = \$100,000$.

Rewrite the following phrase as an algebraic expression. Do not solve.

A carpenter needs 15 nails for every foot of lumber he uses. If he purchases 130 feet of lumber, how many nails will he need?

Step 1 is to represent the unknown quantity. In this problem, the unknown quantity is the number of nails.

Let x = the total number of nails

Step 2 is to write the left side of the equation. It is known that 15 nails are needed for each foot of lumber. Therefore, the left side of the equation will be:

$(15)(130)$

Step 3 is to write both sides of the equation together.

$(15)(130) = x$

The correct answer is $(15)(130) = x$.

Rewrite the following phrase as an algebraic expression. Do not solve.

A meteorologist measures the air temperature. The temperature was measured to be 68 degrees Fahrenheit. However, he needs to tell the temperature to a European who only understands the temperature in degrees Celsius. If the temperature in Celsius multiplied by 1.8 plus 32 equals the temperature in Fahrenheit, what is the temperature in Celsius?

Step 1 is to represent the unknown quantity. In this problem, the unknown quantity is the temperature in degrees Celsius.

Let x = temperature in degrees Celsius

Step 2 is to write the left side of the equation. It is known that the temperature in Celsius multiplied by 1.8 plus 32 equals the temperature in Fahrenheit. Therefore, the left side of the equation will be:

$1.8x + 32$

Step 3 is to write the right side of the equation. It is known that the temperature was measured to be 68 degrees Fahrenheit. Therefore, the right side of the equation will be:

68

Step 4 is to write both sides of the equation together.

$1.8x + 32 = 68$

The correct answer is $1.8x + 32 = 68$.

Q Rewrite the following phrase as an algebraic expression. Put the equation into the standard quadratic equation. Do not solve.

A chemist is developing a chemical formula that will remove paint from metal. The formula requires that the amount of oxygen (in grams) plus 2 equals the amount of oxygen squared. How much oxygen is needed?

 Step 1 is to represent the unknown quantity. In this problem, the unknown quantity is the amount of oxygen.

Let x = the amount of oxygen

Step 2 is to write the left side of the equation. It is known that the amount of oxygen (in grams) plus 2 equals the amount of oxygen squared. Therefore, the left side of the equation will be:

$x + 2$

Step 3 is to write the right side of the equation. It i: the result will be the amount of oxygen squared. Therefore, $\ldots$ right side of the equation will be:

$$x^2$$

Step 4 is to write both sides of the equation together.

$$x + 2 = x^2$$

Step 5 is to put the equation into the standard quadratic equation by subtracting x and 2 from both sides.

$$x + 2 - x - 2 = x^2 - x - 2$$

Step 6 is to simplify the equation.

$$x^2 - x - 2 = 0$$

The correct answer is $x^2 - x - 2 = 0$.

 Rewrite the following phrase as an algebraic expression. Do not solve.

A truck driver needs 1 gallon of gas for every 12 miles that she drives. If the driver is traveling at 50 miles per hour (m.p.h.) and uses 10 gallons of gas, how far did she drive?

 Step 1 is to represent the unknown quantity. In this problem, the unknown quantity is distance travelled by the driver.

Let x = distance traveled by the driver

Step 2 is to write the left side of the equation. It is known that the truck driver needs 1 gallon of gas for every 12 miles that she drives. The problem also states that the driver is traveling at 50 miles per hour (m.p.h.). This is extraneous information and is not used in the equation. Therefore, the left side of the equation will be:

$$\frac{x}{12}$$

Step 3 is to write the right side of the equation. It is known that 10 gallons of gas is used. Therefore, the right side of the equation will be:

10

Step 4 is to write both sides of the equation together.

$$\frac{x}{12} = 10$$

The correct answer is $\frac{x}{12} = 10$.

 Rewrite the following phrase as an algebraic expression. Do not solve.

A tug boat's average speed is 10 miles per hour. If the tug boat travels for 14 hours and the distance it covers is $\frac{1}{2}$ the distance a speed boat could cover, what is the average speed of the speed boat?

 Step 1 is to represent the unknown quantity. In this problem, the unknown quantity is the average speed of the speed boat.

Let x = the average speed of the speed boat

Step 2 is to write the left side of the equation. It is known that the tug boat's average speed is 10 miles per hour and the tug boat travels for 14 hours. It is also known that the distance it covers is $\frac{1}{2}$ the distance a speed boat covers. Therefore, the left side of the equation will be:

$$\left(\frac{14 \text{ hours} \times 10 \text{ miles}}{\text{hour}} \right)$$

Step 3 is to write the right side of the equation. It is known that the distance it covers is $\frac{1}{2}$ the distance a speed boat could cover. Therefore, the right side of the equation will be:

$$\frac{x}{2}$$

Step 4 is to write both sides of the equation together.

$$\left(\frac{14 \text{ hours} \times 10 \text{ miles}}{\text{hour}} \right) = \left(\frac{x}{2} \right)$$

The correct answer is $\left(\dfrac{14 \text{ hours} \times 10 \text{ miles}}{\text{hour}} \right) = \left(\dfrac{x}{2} \right)$ hours.

10.3 Problem Solving with Algebraic Equations

 Rewrite the following phrase as an algebraic expression. Solve the expression.

Ten is subtracted from a number. It is then divided by 5. The quotient equals 2. What is the number?

 Step 1 is to represent the unknown quantity. In this problem, the unknown quantity is the missing number.

Let x = the missing number

Step 2 is to write the left side of the equation. It is known that the missing number minus 10, then divided by 5, will equal the given quotient. Therefore, the left side of the equation will be:

$$\frac{(x-10)}{5}$$

Step 3 is to write the right side of the equation. It is known that the quotient is 2. Therefore, the right side of the equation will be:

$$2$$

Step 4 is to write both sides of the equation together.

$$\frac{(x-10)}{5} = 2$$

Step 5 is to solve the equation.

$$\frac{(x-10)}{5} = 2$$

$$x - 10 = 10$$

$$x = 20$$

The correct answer is $x = 20$.

 Rewrite the following phrase as an algebraic expression. Solve the expression.

Kelly gave $25 to her cousin Kayde for her birthday. Combined with her savings, Kayde now has $700 toward college. How much does Kayde have in savings?

 Step 1 is to represent the unknown quantity. In this problem, the unknown quantity is the amount of money that Kayde has in savings.

Let x = the amount of money in Kayde's savings

Step 2 is to write the left side of the equation. It is known that the amount Kayde received for her birthday plus the amount she has saved will equal the total amount. Therefore, the left side of the equation will be:

$25 + x$

Step 3 is to write the right side of the equation. It is known that the total amount is $700. Therefore, the right side of the equation will be:

$700

Step 4 is to write both sides of the equation together.

$25 + x = $700

Step 5 is to solve the equation.

$25 + x = $700

$$x = $700 - $25$$

$$x = $675$$

The correct answer is $x = 675.

 Rewrite the following phrase as an algebraic expression. Solve the expression.

Island Beach Inc. is a real estate developer. This year's house sales will be $\frac{1}{2}$ as much as last year's sales plus 5% (of last year's sales). If this year's sales is 4,100 homes, how much was last year's sales?

Step 1 is to represent the unknown quantity. In this problem, the unknown quantity is last year's sales.

Let x = last year's sales

Step 2 is to write the left side of the equation. It is known that last year's sales times $\frac{1}{2}$ plus 5% will equal this year's sales. Therefore, the left side of the equation will be:

$$\frac{1}{2}x + 5\%x$$

Step 3 is to write the right side of the equation. It is known that this year's sales is 4,100 homes. Therefore, the right side of the equation will be:

4,100

Step 4 is to write both sides of the equation together.

$$\frac{1}{2}x + 5\%x = 4,100$$

Step 5 is to solve the equation.

$$\frac{1}{2}x + 5\%x = 4,100$$

$$\frac{1}{2}x + 0.05x = 4,100$$

$$0.55x = 4,100$$

$$x = 7,455$$

The correct answer is $x = 7,455$ homes.

Rewrite the following phrase as an algebraic expression. Solve the expression.

A painter needs 4 brushes for every picture he paints. If he paints 52 pictures a year, assuming he doesn't clean out his brushes and reuse them, how many brushes will he need?

 Step 1 is to represent the unknown quantity. In this problem, the unknown quantity is the number of brushes.

Let x = number of brushes per year

Step 2 is to write the left side of the equation. It is known that 4 brushes are needed for each picture and he paints 52 pictures a year. Therefore, the left side of the equation will be:

4(52)

Step 3 is to write the right side of the equation. The right side of the equation is x, the number of brushes per year.

x

Step 4 is to write both sides of the equation together.

$4(52) = x$

Step 5 is to solve the equation.

$4(52) = x$

$208 = x$

The correct answer is $x = 208$ brushes.

 Rewrite the following phrase as an algebraic expression. Solve the expression.

A land surveyor measured the distance between the Price County Court House and Lake Silver. The distance was measured to be

6,320 meters. However, he needs to convert the distance to kilometers (1 km = 1,000 m). He then must subtract a flood factor (the lake rises during the rainy season). If the distance in meters divided by 1,000 minus the flood factor equals 5.25 kilometers, what is the flood factor?

 Step 1 is to represent the unknown quantity. In this problem, the unknown quantity is the flood factor in kilometers.

Let x = the flood factor in kilometers

Step 2 is to write the left side of the equation. It is known that the distance was measured to be 6,320 meters minus the flood factor. Therefore, the left side of the equation will be:

$$\frac{6,320}{1,000} - x$$

Step 3 is to write the right side of the equation. It is known that the total distance is 5.25 kilometers. Therefore, the right side of the equation will be:

5.25 kilometers

Step 4 is to write both sides of the equation together.

$$\frac{6,320}{1,000} - x = 5.25$$

Step 5 is to solve the equation.

$$6.32 - x = 5.25$$
$$-x = -1.07$$
$$x = 1.07$$

The correct answer is $x = 1.07$ kilometers.

Rewrite the following phrase as an algebraic expression. Solve the expression.

A number squared plus 6 times the number equals –8.

 Step 1 is to represent the unknown quantity. In this problem, the unknown quantity is the missing number.

Let x = the missing number

Step 2 is to write the left side of the equation. It is known that the missing number squared, plus 6 times the missing number, equals –8. Therefore, the left side of the equation will be:

$$x^2 + 6x$$

Step 3 is to write the right side of the equation. It is known that the result will be –8. Therefore, the right side of the equation will be:

$$-8$$

Step 4 is to write both sides of the equation together.

$$x^2 + 6x = -8$$

Step 5 is to put the equation into the standard quadratic equation, by adding 8 to both sides.

$$x^2 + 6x + 8 = -8 + 8$$

Step 6 is to simplify the equation.

$$x^2 + 6x + 8 = 0$$

Step 7 is to solve the equation.

$$x^2 + 6x + 8 = 0$$

$$(x + 4)(x + 2) = 0$$

$$x = -4, \quad x = -2$$

The correct answers are $x = -4$ or $x = -2$.

Rewrite the following phrase as an algebraic expression. Solve the expression.

An engineer is developing a theory on nuclear reactions. The theory states that 4 times the amount of plutonium (in grams) needed plus 5.00 equals the amount of plutonium squared. How much plutonium is needed?

Step 1 is to represent the unknown quantity. In this problem, the unknown quantity is the amount of plutonium.

Let x = the amount of plutonium

Step 2 is to write the left side of the equation. It is known that 4 times the amount of plutonium (in grams) plus 5.00 equals the amount of plutonium squared. Therefore, the left side of the equation will be:

$$4x + 5.00$$

Step 3 is to write the right side of the equation. It is known that the result will be the amount of plutonium squared. Therefore, the right side of the equation will be:

$$x^2$$

Step 4 is to write both sides of the equation together.

$$4x + 5 = x^2$$

Step 5 is to put the equation into the standard quadratic equation, by subtracting $4x$ and 5 from both sides.

$$4x + 5 - 4x - 5 = x^2 - 4x - 5$$

Step 6 is to simplify the equation.

$$x^2 - 4x - 5 = 0$$

Step 7 is to solve the equation.

$$x^2 - 4x - 5 = 0$$

$$(x - 5)(x + 1) = 0$$

$$x = 5 \qquad x = -1$$

Since −1 grams is not a valid value, the correct answer is 5 grams of plutonium.

 Rewrite the following phrase as an algebraic expression. Solve the expression.

Farmer Ted's cows produce 1 gallon of milk for every 6 pounds of feed they consume. If the average weight of a cow is 1,500 pounds and produces 10 gallons of milk, how much feed will the cow consume?

Step 1 is to represent the unknown quantity. In this problem, the unknown quantity is the amount of feed consumed by the cow.

Let x = the amount of feed consumed by the cow

Step 2 is to write the left side of the equation. It is known that the cows produce 1 gallon of milk for every 6 pounds of feed they consume. It is also known that the average weight of a cow is 1,500 pounds. This is extraneous information and is not used in the equation. Therefore, the left side of the equation will be:

$$\frac{x}{6}$$

Step 3 is to write the right side of the equation. It is known that 10 gallons of milk are produced. Therefore, the right side of the equation will be:

10

Step 4 is to write both sides of the equation together.

$$\frac{x}{6} = 10$$

Step 5 is to solve the expression.

$$\frac{x}{6} = 10$$

$$x = 10(6)$$

$$x = 60$$

The correct answer is $x = 60$ pounds of feed.

 Rewrite the following phrase as an algebraic expression. Solve the expression.

A jet's average speed is 320 miles per hour. If the jet travels for 2 hours and the distance it covers is 2 times the distance a propeller plane could cover in 2 hours, what is the average speed of the propeller plane?

 Step 1 is to represent the unknown quantity. In this problem, the unknown quantity is the average speed of the propeller plane.

Let $x =$ the average speed of the propeller plane

Step 2 is to write the left side of the equation. It is known that the jet's average speed is 320 miles per hour and it travels for 2 hours. Therefore, the left side of the equation will be:

$$\left(\frac{2 \text{ hours} \times 320 \text{ miles}}{\text{hour}} \right)$$

Step 3 is to write the right side of the equation. It is known that the distance it covers is 2 times the distance of a propeller plane. Therefore, the right side of the equation will be:

$2x$

Step 4 is to write both sides of the equation together.

$$\left(\frac{2\text{ hours } \times 320\text{ miles}}{\text{hour}}\right) = 2x$$

Step 5 is to solve the expression.

640 miles $= 2x$

$x = 320$ miles

In step 6, since the propeller plane travels 320 miles in 2 hours, its speed is:

$$\frac{320}{2}\text{ hours} = 160\text{ miles per hour}$$

The correct answer is 160 miles per hour.

 Rewrite the following phrase as an algebraic expression. Solve the expression.

When Jake's dad turns 60, he will be 4 times Jake's current age. What is Jake's current age?

 Step 1 is to represent the unknown quantity. In this problem, the unknown quantity is Jake's current age.

Let $x =$ Jake's current age

Step 2 is to write the left side of the equation. It is known that Jake's current age times 4 will equal his dad s age in the future. Therefore, the left side of the equation will be:

4x

Step 3 is to write the right side of the equation. It is known that his dad's age will be 60. Therefore, the right side of the equation will be:

60

Step 4 is to write both sides of the equation together.

$4(x) = 60$

Step 5 is to solve the equation.

$4x = 60$

$x = 15$

The correct answer is that Jake's current age is 15.

10.4 Word Problems with Inequalities

 Rewrite the following phrase as an algebraic expression. Do not solve.

A number subtracted from 8 is less than 20. What is the number?

 Step 1 is to represent the unknown quantity.

Let x = the unknown number

Step 2 is to set up the left side of the inequality. It is known that the unknown number is subtracted from 8. Therefore, the left side of the inequality will be:

$8 - x$

Step 3 is to set up the right side of the inequality. It is known that the unknown number, when subtracted from 8, will be less than 20. Therefore, the right side of the inequality will be:

20

Step 4 is to write both sides of the inequality together. Remember to use the "less than" sign.

$8 - x < 20$

The correct answer is $8 - x < 20$.

 Rewrite the following phrase as an algebraic expression. Do not solve.

Ms. Taylor is a traveling salesperson. She must sell at least $200 per week to stay employed with her company, Vacuums R Us. If each unit she sells is $58, how many units must she sell to stay employed?

 Step 1 is to represent the unknown quantity. In this problem, the unknown quantity is the number of units sold.

Let x = the number of units sold

Step 2 is to set up the left side of the inequality. It is known that each unit sold is $58. Therefore, the left side of the inequality will be:

$58x

Step 3 is to set up the right side of the inequality. It is known that she must sell at least $200 worth of product to stay employed. Therefore, the right side of the inequality will be:

$200

Step 4 is to write both sides of the inequality together. Remember to use the "greater than or equal to" sign.

$58x \geq \$200$

The correct answer is $58x \geq \$200$.

 Rewrite the following phrase as an algebraic expression. Solve the expression.

A number, when divided by 2, is less than 2 times the number plus 2. What is the number?

 Step 1 is to represent the unknown quantity.

Let x = the unknown number

Step 2 is to set up the left side of the inequality. It is known that the unknown number is divided by 2. Therefore, the left side of the inequality will be:

$$\frac{x}{2}$$

Step 3 is to set up the right side of the inequality. It is known that the unknown number, when divided by 2, will be less than 2 times the number plus 2. Therefore, the right side of the inequality will be:

$$2x + 2$$

Step 4 is to write both sides of the inequality together. Remember to use the "less than" sign.

$$\frac{x}{2} < 2x + 2$$

Step 5 is to solve the expression.

$$\frac{x}{2} < 2x + 2$$

$$x < 4x + 4$$

$$-3x < 4$$

$$-x < \frac{4}{3}$$

$$x > -\frac{4}{3}$$

The correct answer is $x > -\dfrac{4}{3}$.

 Rewrite the following phrase as an algebraic expression. Solve the expression.

A stock broker has a formula for making the most profit for his customer. For each share of the Desert Technology Corporation, his customer will make $3 profit plus $.50 for each share from dividends. If his customer wants to make at least $3,500 profit, how many shares are needed?

 Step 1 is to represent the unknown quantity. In this problem, the unknown quantity is the number of shares of stock.

Let x = the number of shares of stock

Step 2 is to set up the left side of the inequality. It is known that each share brings in a profit of $3 plus an extra $.50 from dividends. Therefore, the left side of the inequality will be:

$3x + $0.50x$

Step 3 is to set up the right side of the inequality. It is known that he must make at least $3,500 profit. Therefore, the right side of the inequality will be:

$3,500

Step 4 is to write both sides of the inequality together. Remember to use the "greater than or equal to" sign.

$3x + $.50x \geq $3,500

Step 5 is to solve the expression.

$3x + $.50x \geq $3,500

$3.5x \geq $3,500

$$x \geq 1,000$$

The correct answer is $x \geq 1,000$. The stock broker has to buy at least 1,000 shares of the stock to make $3,500 profit for his customer.

 Rewrite the following phrase as an algebraic expression. Solve the expression.

A number squared plus 10 is less than or equal to 154. What is the number?

 Step 1 is to represent the unknown quantity.

Let x = the unknown number

Step 2 is to set up the left side of the inequality. It is known that the unknown number is squared and then 10 is added. Therefore, the left side of the inequality will be:

$x^2 + 10$

Step 3 is to set up the right side of the inequality. It is known that the unknown number, when squared and then added to 10, will be less than or equal to 154. Therefore, the right side of the inequality will be:

154

Step 4 is to write both sides of the inequality together. Remember to use the "less than or equal to" sign.

$x^2 + 10 \le 154$

Step 5 is to solve the expression.

$x^2 + 10 \le 154$

$x^2 \le 144$

$x \le 12$ and $x \ge -12$

The correct answer is $-12 \le x \le 12$.

CHAPTER 11

Graphing and
Coordinate Geometry

Coordinate geometry refers to the study of geometric figures using algebraic principles.

The graph shown is called the Cartesian coordinate plane. The graph consists of a pair of perpendicular lines called **coordinate axes**. The **vertical axis** is the y-axis and the **horizontal axis** is the x-axis. The point of intersection of these two axes is called the **origin**; it is the zero point of both axes. Furthermore, points to the right of the origin on the x-axis and above the origin on the y-axis represent positive real numbers. Points to the left of the origin on the x-axis or below the origin on the y-axis represent negative real numbers.

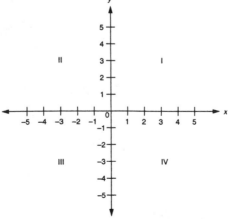

The four regions cut off by the coordinate axes are, in counter-clockwise direction from the top right, called the first, second, third and fourth quadrant, respectively. The first quadrant contains all points with two positive coordinates.

In the graph shown, two points are identified by the ordered pair, (x, y) of numbers. The x-coordinate is the first number and the y-coordinate is the second number.

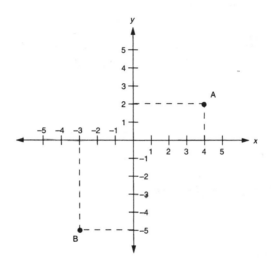

To plot a point on the graph when given the coordinates, draw perpendicular lines from the number-line coordinates to the point where the two lines intersect.

To find the coordinates of a given point on the graph, draw perpendicular lines from the point to the coordinates on the number line. The x-coordinate is written before the y-coordinate and a comma is used to separate the two.

In this case, point A has the coordinates $(4, 2)$ and the coordinates of point B are $(-3, -5)$.

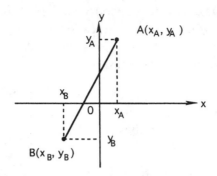

For any two points A and B with coordinates (X_A, Y_A) and (X_B, Y_B), respectively, the distance between A and B is represented by:

$$AB = \sqrt{(X_A - X_B)^2 + (Y_A - Y_B)^2}$$

This is commonly known as the distance formula or the **Pythagorean Theorem**.

11.1   Plotting Points

 Plot the point (–3, 2) on the graph below.

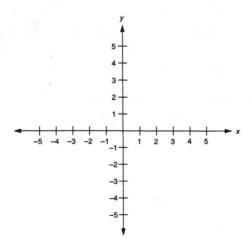

Step 1 is to find the point −3 along the *x*-axis. Draw a dashed vertical line through that point. See the figure below.

Step 2 is to find the point 2 along the *y*-axis. Draw a dashed horizontal line through that point. See the figure below.

Step 3 is to plot the point where the two dashed lines intersect with a solid circle. This is the point (−3, 2). See the figure below.

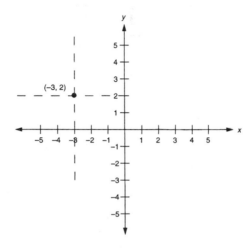

 Plot the point (4, −4) on the graph below.

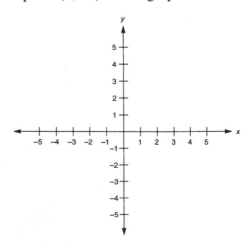

Step 1 is to find the point 4 along the *x*-axis. Draw a dashed vertical line through that point. See the figure below.

Step 2 is to find the point −4 along the *y*-axis. Draw a dashed horizontal line through that point. See the figure below.

Step 3 is to plot the point where the two dashed lines intersect with a solid circle. This is the point (4, −4). See the figure below.

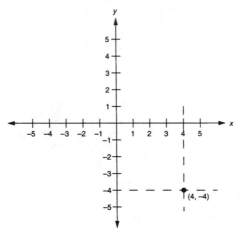

Plot the point (−20, −60) on the graph below.

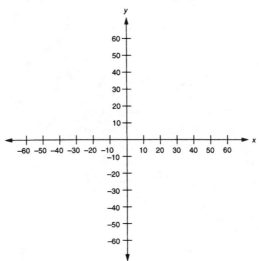

Step 1 is to find the point –20 along the *x*-axis. Draw a dashed vertical line through that point. See the figure below.

Step 2 is to find the point –60 along the *y*-axis. Draw a dashed horizontal line through that point. See the figure below.

Step 3 is to plot the point where the two dashed lines intersect with a solid circle. This is the point (–20, –60). See the figure below.

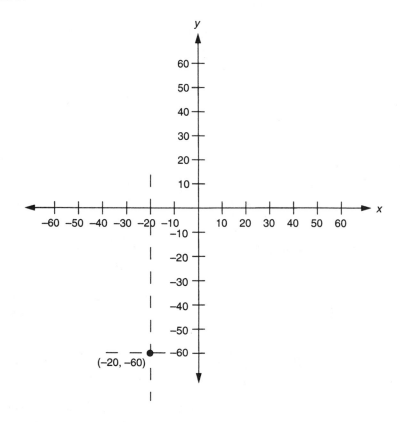

Plot the points (2, 5) and (5, –5) on the graph below. Draw a line connecting the points.

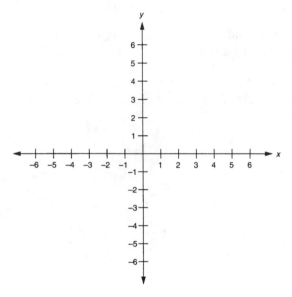

 Step 1 is to find the point 2 along the *x*-axis. Draw a dashed vertical line through that point. See the figure on the next page.

Step 2 is to find the point 5 along the *y*-axis. Draw a dashed horizontal line through that point. See the figure on the next page.

Step 3 is to plot the point where the two dashed lines intersect with a solid circle. This is the point (2, 5). See the figure on the next page.

Step 4 is to find the point 5 along the *x*-axis. Draw a dashed vertical line through that point. See the figure on the next page.

Step 5 is to find the point –5 along the *y*-axis. Draw a dashed horizontal line through that point. See the figure on the next page.

Step 6 is to plot the point where the two dashed lines intersect with a solid circle. This is the point (5, –5). See the figure below.

Step 7 is to draw a straight line connecting the two points.

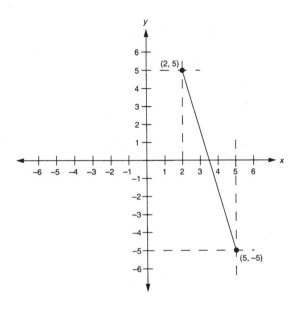

 Graph the following points on a graph. Draw and label the x- and y-axes.

(10, 5) (–30, –10) (–5, –35)

A When drawing the x- and y-axes, label the axis incorporating coordinates that are used in the points. To do this, find the x- and y-coordinates, then plot the intersection of the coordinates as in the figure. See the figure on the next page for an example of a graph.

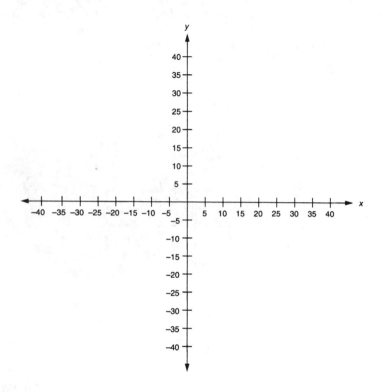

See the figure on the next page for the correct graph with the plotted points.

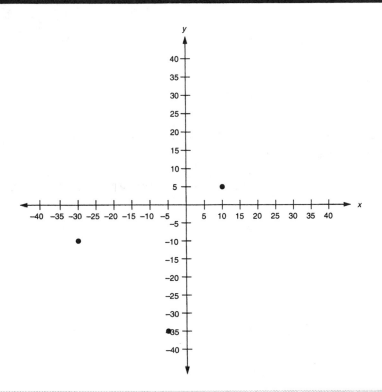

11.2 Slope

 A line contains the points (3, 1) and (5, 6). What is the slope of the line?

 Step 1 is to use the coordinates in the formula for the slope of a line. The formula is $\dfrac{(y_2 - y_1)}{(x_2 - x_1)}$.

$$\text{slope} = \frac{6-1}{5-3}$$

Step 2 is to solve the formula to obtain the slope.

$$\text{slope} = \frac{6-1}{5-3} = \frac{5}{2}$$

The slope of the line is $\frac{5}{2}$.

 A line contains the points (–6, 3) and (–3, 6). What is the slope of the line?

 Step 1 is to use the coordinates in the formula for the slope

of a line. The formula is $\frac{(y_2 - y_1)}{(x_2 - x_1)}$.

$$\text{slope} = \frac{6-3}{-3-(-6)}$$

Step 2 is to solve the formula to obtain the slope.

$$\text{slope} = \frac{6-3}{-3+6} = \frac{3}{3} = 1$$

The slope of the line is 1.

 Line a contains the points (–1, 4) and (4, 6). Line b contains the points (5, 2) and (–3, –3). Are the two lines parallel?

 Step 1 is to calculate the slope for Line a. The formula for

the slope of a line is $\frac{(y_2 - y_1)}{(x_2 - x_1)}$.

$$\frac{6-4}{4-(-1)} = \frac{2}{5}$$

Step 2 is to calculate the slope for Line b. The formula for the slope of a line is $\dfrac{(y_2 - y_1)}{(x_2 - x_1)}$.

$$\frac{-3-2}{-3-5} = \frac{-5}{-8} = \frac{5}{8}$$

Parallel lines have the same slope. Since the slope of Line a does not equal the slope of Line b, the lines cannot be parallel.

 Line a contains the points (2, 4) and (6, 8). Line b contains the points (4, 8) and (12, 16). Are the two lines parallel?

 Step 1 is to calculate the slope for Line a. The formula for the slope of a line is $\dfrac{(y_2 - y_1)}{(x_2 - x_1)}$.

$$\frac{(8-4)}{(6-2)} = \frac{4}{4} = 1$$

Step 2 is to calculate the slope for Line b. The formula for the slope of a line is $\dfrac{(y_2 - y_1)}{(x_2 - x_1)}$.

$$\frac{(16-8)}{(12-4)} = \frac{8}{8} = 1$$

Parallel lines have the same slope. Since the slope of Line a equals the slope of Line b, the lines must be parallel.

Line a contains the points (–2, –2) and (6, 4). Line b contains the points (0, 4) and (3, 0). Are the two lines perpendicular?

 Step 1 is to calculate the slope for Line a. The formula for the slope of a line is $\dfrac{(y_2 - y_1)}{(x_2 - x_1)}$.

$$\frac{4-(-2)}{6-(-2)} = \frac{4+2}{6+2} = \frac{6}{8} = \frac{3}{4}$$

Step 2 is to calculate the slope for Line b. The formula for the slope of a line is $\dfrac{(y_2 - y_1)}{(x_2 - x_1)}$.

$$\frac{0-4}{3-0} = -\frac{4}{3}$$

Two lines are perpendicular if the slope of one line (m) equals the negative reciprocal of the other $\left(\dfrac{-1}{m}\right)$. Since the slope of Line a is $\dfrac{3}{4}$ and the slope of Line b is $\dfrac{-4}{3}$, the lines are perpendicular.

Use the graph below to calculate the slope of Line a.

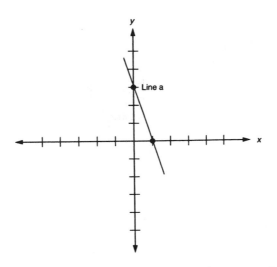

Line a

 Step 1 is to find two points that are intersected by Line a.

(1, 0) and (0, 3)

Step 2 is to use the formula for the slope of a line to calculate

the slope of Line a. The formula is $\dfrac{(y_2 - y_1)}{(x_2 - x_1)}$.

$$\text{slope} = \frac{3-0}{0-1}$$

Step 3 is to solve the formula to obtain the slope.

$$\text{slope} = \frac{3-0}{0-1} = -\frac{3}{1} = -3$$

The slope of Line a is −3.

Use the graph below to calculate the slope of Line a.

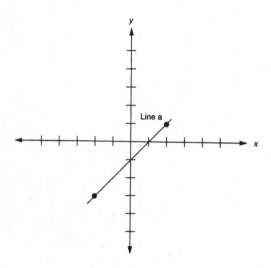

 Step 1 is to find two points that are intersected by Line a.

(2, 1) and (–2, –3)

Step 2 is to use the formula for the slope of a line to calculate the slope of Line a. The formula is $\dfrac{(y_2 - y_1)}{(x_2 - x_1)}$.

$$\text{slope} = \frac{-3-1}{-2-2}$$

Step 3 is to solve the formula to obtain the slope.

$$\text{slope} = \frac{-3-1}{-2-2} = \frac{-4}{-4} = 1$$

The slope of Line a is 1.

11.3 Graphing Linear Equations

 Find three pairs of coordinates that the line below passes through.

$$y = 3x + 7$$

 Step 1 is to pick three arbitrary values for x.

0, 1, 2

Step 2 is to substitute each value for x in the equation to obtain y. Make a chart to log the values.

x	y
0	7
1	10
2	13

$$y = 3(0) + 7 = 7$$

$$y = 3(1) + 7 = 10$$

$$y = 3(2) + 7 = 13$$

Step 3 is to rewrite each set of coordinates.

(0, 7), (1, 10), (2, 13)

The correct answers are (0, 7), (1, 10), and (2, 13). However, there are many possible answers. Any answer is valid if the coordinates make the equation true.

Find three pairs of coordinates that the line below passes through.

$$y = \frac{1}{4}x - 2$$

 Step 1 is to pick three arbitrary values for x.

0, 4, 8

Step 2 is to substitute each value for x in the equation to obtain y. Make a chart to log the values.

x	y
0	−2
4	−1
8	0

$$y = \frac{1}{4}(0) - 2 = -2$$

$$y = \frac{1}{4}(4) - 2 = -1$$

$$y = \frac{1}{4}(8) - 2 = 0$$

Step 3 is to rewrite each set of coordinates.

(0, −2), (4, −1), (8, 0)

The correct answers are (0, −2), (4, −1), and (8, 0). However, there are many possible answers. Any answer is valid if the coordinates make the equation true.

Put the following equation in y slope–intercept form.

$$\frac{1}{3}y - \frac{1}{6}x = 2$$

 Step 1 is to isolate the variable $\frac{1}{3}y$. To do this, add $\frac{1}{6}x$ to both sides of the equation.

$$\frac{1}{3}y - \frac{1}{6}x + \frac{1}{6}x = 2 + \frac{1}{6}x$$

Step 2 is to simplify the left side of the equation.

$$\frac{1}{3}y - \frac{1}{6}x + \frac{1}{6}x = \frac{1}{3}y$$

Step 3 is to simplify the right side of the equation.

$$2 + \frac{1}{6}x = 2 + \frac{1}{6}x$$

Step 4 is to rewrite the equation.

$$\frac{1}{3}y = 2 + \frac{1}{6}x$$

Step 5 is to multiply both sides of the equation by 3 to isolate the variable y.

$$\frac{1}{3}y(3) = 3\left(2 + \frac{1}{6}x\right)$$

$$y = 6 + \frac{1}{2}x$$

Step 6 is to rewrite the equation to be in the form $y = mx + b$, where m is the slope and b is the y-intercept.

$$y = \frac{1}{2}x + 6$$

The correct answer is $y = \frac{1}{2}x + 6$.

Find the slope of the following equation.

$$\frac{1}{2}y - \frac{1}{4}x - 4 = 0$$

Step 1 is isolate the variable $\frac{1}{2}y$.

$$\frac{1}{2}y - \frac{1}{4}x - 4 = 0$$

$$\frac{1}{2}y = \frac{1}{4}x + 4$$

Step 2 is to multiply both sides of the equation by 2. This will isolate the variable y and put the equation in y slope–intercept form.

$$y = \frac{1}{2}x + 8$$

Step 3 is to find the slope of the line. The coefficient of x is the slope.

$$m = \frac{1}{2}$$

The correct answer is $m = \dfrac{1}{2}$.

 Write an equation of a line with the slope m and the y intercept b.

$$\begin{cases} \text{when} \quad m = -\dfrac{1}{4} \\ \text{and} \quad b = 11 \end{cases}$$

 Step 1 is to write the standard form of an equation in y slope–intercept form.

$y = mx + b$

Step 2 is to substitute m into the equation.

$$y = -\dfrac{1}{4}x + b$$

Step 3 is to substitute b into the equation.

$$y = -\dfrac{1}{4}x + 11$$

The correct answer is $y = -\dfrac{1}{4}x + 11$.

 Graph the equation $\dfrac{1}{2}y = x - 1$.

 Step 1 is to put the equation in y slope–intercept form.

$$\dfrac{1}{2}y = x - 1$$

$$y = 2x - 2$$

Step 2 is to choose two arbitrary values of x. Log the values in a chart. Use the values of x to obtain values for y.

x	y
0	–2
3	4

Step 3 is to use the x and y values to form coordinates. Note that there are many different possible coordinates. However, the graph should look the same.

(0, –2) and (3, 4)

Step 4 is to use the coordinates to graph the line.

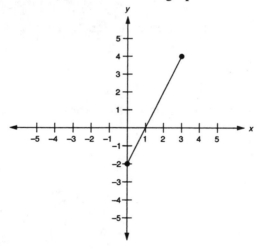

 Graph the equation $\dfrac{1}{4}y = \dfrac{1}{2}x - 1$.

 Step 1 is to put the equation in y slope–intercept form.

$$\frac{1}{4}y = \frac{1}{2}x - 1$$

$$y = 2x - 4$$

Step 2 is to choose two arbitrary values of x. Log the values in a chart. Use the values of x to obtain values for y.

x	y
0	–4
2	0

Step 3 is to use the x and y values to form coordinates. Note that there are many different possible coordinates. However, the graph should look the same.

(0, –4) and (2, 0)

Step 4 is to use the coordinates to graph the line.

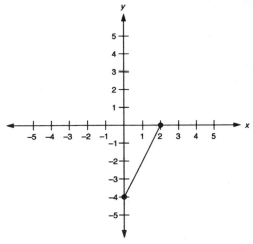

11.4   Graphing Inequalities

 Put the following inequality in y slope–intercept form.

$$y + 5x > 22$$

Step 1 is to isolate the variable y. To do this, subtract $5x$ from both sides of the inequality.

$$y + 5x - 5x > 22 - 5x$$

Step 2 is to simplify the left side of the inequality.

$$y + 5x - 5x = y$$

Step 3 is to simplify the right side of the inequality.

$$22 - 5x = 22 - 5x$$

Step 4 is to rewrite the inequality.

$$y > 22 - 5x$$

Step 5 is to rewrite the inequality to be in the form $y = mx + b$, where m is the slope and b is the y–intercept.

$$y > -5x + 22$$

The correct answer is $y > -5x + 22$.

Put the following inequality in y slope–intercept form.

$$\frac{1}{8}y - \frac{1}{4}x < 2$$

Step 1 is to isolate the variable $\frac{1}{8}y$. To do this, add $\frac{1}{4}x$ to both sides of the inequality.

$$\frac{1}{8}y - \frac{1}{4}x + \frac{1}{4}x < 2 + \frac{1}{4}x$$

Step 2 is to simplify the left side of the inequality.

$$\frac{1}{8}y - \frac{1}{4}x + \frac{1}{4}x = \frac{1}{8}y$$

Step 3 is to simplify the right side of the inequality.

$$2 + \frac{1}{4}x = 2 + \frac{1}{4}x$$

Step 4 is to rewrite the inequality.

$$\frac{1}{8}y < 2 + \frac{1}{4}x$$

Step 5 is to multiply both sides of the inequality by 8 to isolate the variable y.

$$\frac{1}{8}y(8) < 8\left(2 + \frac{1}{4}x\right)$$

$$y < 16 + 2x$$

Step 6 is to rewrite the inequality to be in the form $y = mx + b$, where m is the slope and b is the y–intercept.

$$y < 2x + 16$$

The correct answer is $y < 2x + 16$.

 Graph the inequality $-y > x + 5$.

 Step 1 is to put the inequality in y slope–intercept form.

$$-y > x + 5$$

$$y < -x - 5$$

Step 2 is to choose two arbitrary values of *x*. Log the values in a chart. Use the values of *x* to obtain values for *y*. This will solve the equation $y = -x - 5$.

x	*y*
0	-5
-5	0

Step 3 is to use the *x* and *y* values to form coordinates. Note that there are many different possible coordinates. However, the graph should look the same.

(0, -5) and (-5, 0)

Step 4 is to use the coordinates to graph the line. Notice that a dashed line connects the points. The points along the line do not satisfy the inequality. Only the points in the shaded region satisfy the inequality.

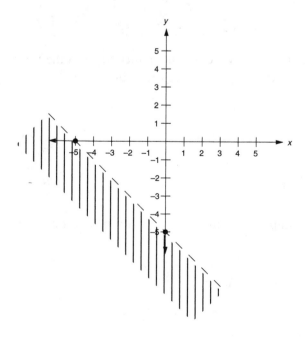

Graph the inequality $4y \leq 4x - 4$.

 Step 1 is to put the inequality in y slope–intercept form.

$4y \leq 4x - 4$

$y \leq x - 1$

Step 2 is to choose two arbitrary values of x. Log the values in a chart. Use the values of x to obtain values for y. These points solve the equation $y = x - 1$.

x	y
0	−1
2	1

Step 3 is to use the x and y values to form coordinates. Note that there are many different possible coordinates. However, the graph should look the same.

(0, −1) and (2, 1)

Step 4 is to use the coordinates to graph the line. Notice that a solid line connects the points. The points along the line satisfy the inequality. The points in the shaded region also satisfy the inequality.

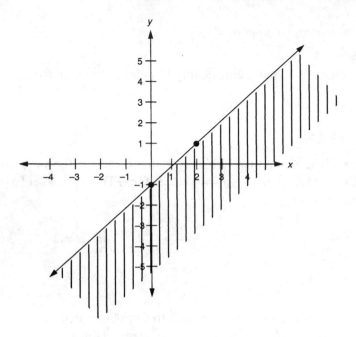

CHAPTER 12

Geometry

12.1 Points, Lines, and Angles

Geometry is built upon a series of undefined terms. These terms are those which we accept as known in order to define other undefined terms.

Point: Although we represent points on paper with small dots, a point has no size, thickness, or width.

Line: A line is a series of adjacent points which extends indefinitely. A line can be either curved or straight; however, unless otherwise stated, the term "line" refers to a straight line.

Plane: A plane is a collection of points lying on a flat surface, which extends indefinitely in all directions.

If A and B are two points on a line, then the **line segment** AB is the set of points on that line between A and B and including A and B, which are endpoints. The line segment is referred to as AB.

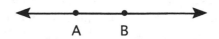

A **ray** is a series of points that lie to one side of a single endpoint.

12.2   Intersection Lines and Angles

An **angle** is a collection of points which is the union of two rays having the same endpoint. An angle such as the one illustrated below can be referred to in any of the following ways:

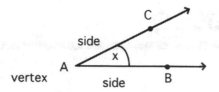

A) by a capital letter which names its vertex, i.e., $\angle A$;

B) by a lowercase letter or number placed inside the angle, i.e., $\angle x$;

C) by three capital letters, where the middle letter is the vertex and the other two letters are not on the same ray, i.e., $\angle CAB$ or $\angle BAC$, both of which represent the angle illustrated in the figure.

12.3   Types of Angles

Vertical angles are formed when two lines intersect. These angles are equal.

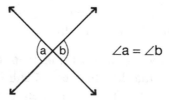

$\angle a = \angle b$

Adjacent angles are two angles with a common vertex and a common side, but no common interior points. In the following figure, $\angle DAC$ and $\angle BAC$ are adjacent angles. $\angle DAB$ and $\angle BAC$ are not.

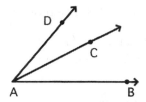

A **right angle** is an angle whose measure is 90°.

An **acute angle** is an angle whose measure is larger than 0° but less than 90°.

An **obtuse angle** is an angle whose measure is larger than 90° but less than 180°.

A **straight angle** is an angle whose measure is 180°. Such an angle is, in fact, a straight line.

A **reflex angle** is an angle whose measure is greater than 180° but less than 360°.

Complementary angles are two angles, the sum of the measures of which equals 90°.

Supplementary angles are two angles, the sum of the measures of which equals 180°.

Congruent angles are angles of equal measure.

12.4 Measuring Angles

 Using a protractor, measure the angle in the figure below.

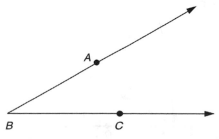

To measure the angle using a protractor follow these steps.

Step 1 is to place the center of the protractor on the vertex of the angle. Side *BC* of the angle should now be aligned with the straight edge of the protractor.

Step 2 is to follow side *AB* from the vertex toward the numbers on the curved edge of the protractor. Read the value where side *AB* crosses the numbers. This is the measure of ∠*ABC*.

The correct answer is 30 degrees.

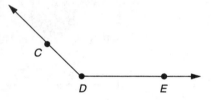 Using a protractor, measure the angle in the figure below.

 To measure the angle using a protractor follow these steps.

Step 1 is to place the center of the protractor on the vertex of the angle. Side *DE* of the angle should now be aligned with the straight edge of the protractor.

Step 2 is to follow side *CD* from the vertex toward the numbers on the curved edge of the protractor. Read the value where side *CD* crosses the numbers. This is the measure of ∠*CDE*.

The correct answer is 135 degrees.

Measure the angles below. Determine if the angles are complementary.

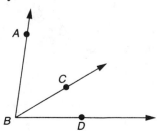

Step 1 is to measure ∠*ABC* by placing the center of the protractor on the vertex of the angle. Side *BC* of the angle should now be aligned with the straight edge of the protractor.

Step 2 is to follow side *AB* from the vertex toward the numbers on the curved edge of the protractor. Read the value where side *AB* crosses the numbers. The measure of ∠*ABC* is 50 degrees.

Step 3 is to measure ∠*CBD* by placing the center of the protractor on the vertex of the angle. Side *BD* of the angle should now be aligned with the straight edge of the protractor.

Step 4 is to follow side *CB* from the vertex toward the numbers on the curved edge of the protractor. Read the value where side *AB* crosses the numbers. The measure of ∠*CBD* is 30 degrees.

Complementary angles are angles that sum to 90 degrees. Since ∠*ABC* and ∠*CBD* do not have a sum of 90, the angles are not complementary.

Q The angles in the figure below are complementary. If ∠*MNO* is 50 degrees, what is the measurement of ∠*ONP*?

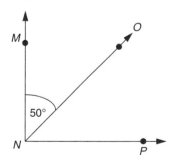

A It is known that the sum of complementary angles equals 90 degrees. If ∠*MNO* equals 50 degrees, then ∠*ONP* equals 90 minus 50.

The correct answer is ∠*ONP* is 40 degrees.

The angles in the figure below are supplementary. If ∠*MNO* is 117 degrees, what is the measurement of ∠*ONP*?

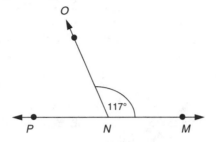

 It is known that the sum of supplementary angles equals 180 degrees. If ∠*MNO* equals 117 degrees, then ∠*ONP* equals 180 minus 117.

The correct answer is ∠*ONP* is 63 degrees.

 Measure the angles below. Determine if the angles are congruent.

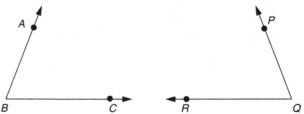

 Step 1 is to measure ∠*ABC* by placing the center of the protractor on the vertex of the angle. Side *BC* of the angle should now be aligned with the straight edge of the protractor.

Step 2 is to follow side *AB* from the vertex toward the numbers on the curved edge of the protractor. Read the value where side *AB* crosses the numbers. The measure of ∠*ABC* is 70 degrees.

Step 3 is to measure ∠*PQR* by placing the center of the protractor on the vertex of the angle. Side *QR* of the angle should now be aligned with the straight edge of the protractor.

Step 4 is to follow side *PQ* from the vertex toward the numbers on the curved edge of the protractor. Read the value where side *PQ* crosses the numbers. The measure of ∠*PQR* is 70 degrees.

Congruent angles are angles that are equal in measurement. Since ∠*ABC* and ∠*PQR* are equal, the angles are congruent.

 Examine the figure below. Which pairs of angles are adjacent?

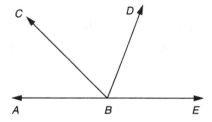

 Adjacent angles are angles that share a common side. The adjacent angles are ∠*ABC* and ∠*CBD*, ∠*ABD* and ∠*DBE*, ∠*CBD* and ∠*DBE*, ∠*ABC* and ∠*CBE*.

12.5 Circles

 Examine the figure below. Locate the following parts of a circle: center, radius, diameter, tangent, secant, chord.

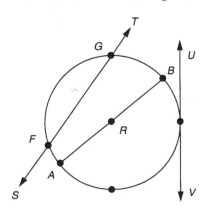

The center of the circle is the point *R*. A circle is defined by its center and all the points on the circle that are an equal distance from the center.

The radius of the circle is the line segment *AR* or *RB*. The radius is defined as the line from the center to any point along the circle.

The diameter of the circle is the line segment *AB*. The diameter is defined as the line with endpoints along the circle and also passes through the center, *R*.

The tangent of the circle is the line *UV*. It is defined as the line that intersects the circle at only one point along the circle.

The secant of the circle is the line *ST*. It is defined as the line that intersects the circle at two points along the circle.

The chord of the circle is the line segment *FG* or *AB*. A chord is defined as a line segment with two endpoints along the circumference of the circle.

 Using the figure below, determine the length of the line segment *CB*.

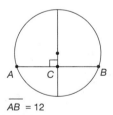

$\overline{AB}$ = 12

 If a chord is perpendicular to the diameter as in the above figure, then the diameter bisects the chord. Since the length of the chord is 12, the line segment *CB* is 6.

Given that the coordinates of the diameter of a circle are (1, 1) and (6, 6), what is the radius of the circle?

Step 1 is to use the distance formula to calculate the diameter.

$$D = \sqrt{(6\text{-}1)^2 + (6\text{-}1)^2}$$
$$D = \sqrt{25+25} = \sqrt{50} = 7.07$$

Step 2 is to divide the diameter by 2 to obtain the radius.

$$r = \frac{7.07}{2}$$
$$r = 3.535$$

The correct answer is $r = 3.535$.

Using the figure below, what is the measure of $\angle ABP$?

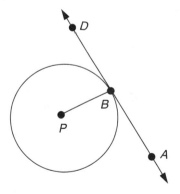

Step 1 is to recognize that line AD is tangent to Circle P.

In Step 2, since line AD is tangent to Circle P, $\angle ABP$ must be a right angle.

The correct answer is 90 degrees.

Examine the figure below. Is line segment AB a chord or the diameter?

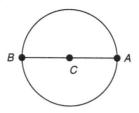

 Line segment AB is both a chord and the diameter. Recall that the definition of a chord is a line segment that has two endpoints on the circumference of a circle. Since all diameters also meet this criteria, diameters are also chords. Also recall that the diameter of a circle is defined as the line with endpoints along the circle and also passes through the center. So, the only chords that are also diameters are those that pass through the center.

 Using the figure below, determine the measure of $\angle ABC$.

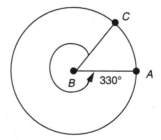

 Since the angle of the arc is given, and a circle is 360 degrees, the measure of $\angle ABC$ must be 360 – the angle of the arc.

$$360 - 330 = 30$$

The correct answer is 30 degrees.

12.6 Triangles

 Determine which of the triangles listed below are right, acute, and obtuse triangles.

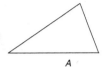 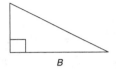 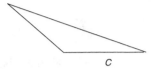

 Triangle *A* is an acute triangle. An acute triangle is a triangle comprised of three acute angles. Triangle *B* is a right triangle. A right triangle is a triangle with a right angle. Triangle *C* is an obtuse triangle. An obtuse triangle is a triangle with one obtuse angle.

 Determine which of the triangles listed below are scalene, isosceles, and equilateral.

  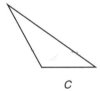

 Triangle *A* is an equilateral triangle. An equilateral triangle is a triangle with three congruent sides. Triangle *B* is an isosceles triangle. An isosceles triangle is a triangle with at least two congruent sides. Therefore, an equilateral triangle is also a type of isosceles triangle. Triangle *C* is a scalene triangle. A scalene triangle is a triangle with no congruent sides.

Find the missing angle in the triangle below.

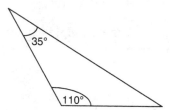

 Since the angles of a triangle always add up to 180 degrees, Step 1 is to total the two known angles.

35 + 110 = 145

Step 2 is to subtract the total in Step 1 from 180.

180 – 145 = 35

The correct answer is 35 degrees.

 Determine if the sides of the triangle below are equal.

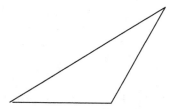

 The triangle is a scalene triangle. Since a scalene triangle does not have any congruent sides, none of the sides of the triangle are equal.

 Determine if the angles of the triangle below are equal.

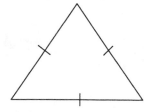

The triangle is equilateral. Since the sides of the triangle are all equal, the angles must be equal too. Therefore, all of the angles in the triangle in this problem are congruent.

 Find the sine of the marked angle in the figure below.

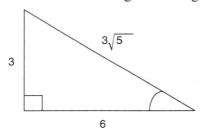

 Step 1 is to write the formula for determining the sine of an angle.

sine = length of opposite side/length of hypotenuse

Step 2 is to substitute the values into the formula.

$$\text{sine} = \frac{3}{3\sqrt{5}} = \frac{1}{\sqrt{5}} = \frac{\sqrt{5}}{5}$$

Step 3 is to calculate the sine.

$$\text{sine} = \frac{\sqrt{5}}{5} \text{ or } 0.447$$

The correct answer is $\frac{\sqrt{5}}{5}$ or 0.447.

 Find the cosine of the marked angle in the figure below.

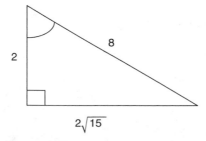

Step 1 is to write the formula for determining the cosine of an angle.

cosine = length of adjacent side/length of hypotenuse

Step 2 is to substitute the values into the formula.

$$\text{cosine} = \frac{2}{8}$$

Step 3 is to calculate the cosine.

cosine = 0.25

The correct answer is 0.25.

 Find the tangent of the marked angle in the figure below.

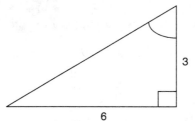

 Step 1 is to write the formula for determining the tangent of an angle.

tangent = length of opposite side/length of adjacent side

Step 2 is to substitute the values into the formula.

$$\text{tangent} = \frac{6}{3}$$

Step 3 is to calculate the tangent.

tangent = 2

The correct answer is 2.

Find the missing angle in the triangle below.

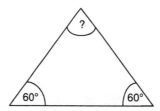

 Since the angles of a triangle always add up to 180 degrees, Step 1 is to total the two known angles.

$60 + 60 = 120$

Step 2 is to subtract the total in Step 1 from 180.

$180 - 120 = 60$

The correct answer is 60 degrees.

 If the sine of $\angle ABC$ in the figure below is $\frac{1}{2}$, what is the length of side AC?

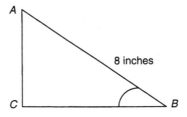

 Step 1 is to write the formula for the sine of an angle.

sine = length of opposite side/length of hypotenuse

Step 2 is to substitute the known values. From the figure it is known that the hypotenuse is 8 inches. It is also known that the sine is $\frac{1}{2}$. The length of the opposite side is x.

$$\frac{1}{2} = \frac{x}{8}$$

Step 3 is to solve the equation.

$x = 4$

The correct answer is that side AC is 4 inches long.

 If the cosine of $\angle ABC$ in the figure below is 0.75, what is the length of side BC?

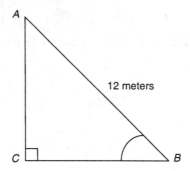

 Step 1 is to write the formula for the cosine of an angle.

cosine = length of adjacent side/length of hypotenuse

Step 2 is to substitute the known values. From the figure it is known that the hypotenuse is 12 meters. It is also known that the cosine is 0.75. The length of the adjacent side is x.

$$0.75 = \frac{x}{12}$$

Step 3 is to solve the equation.

$x = 9$

The correct answer is that side BC is 9 meters long.

12.7 Quadrilaterals

 Identify the quadrilateral in the figure below.

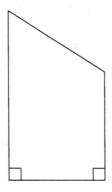

 The quadrilateral is a trapezoid. A trapezoid is a quadrilateral with only one pair of parallel sides.

 Identify the quadrilateral in the figure below.

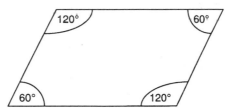

 The quadrilateral is a parallelogram. A parallelogram is a quadrilateral with two pairs of parallel sides.

 Identify the quadrilateral in the figure below.

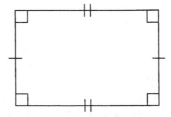

The quadrilateral is a rectangle. A rectangle is a quadrilateral with two pairs of parallel sides and four right angles. A rectangle is also a parallelogram.

 Identify the quadrilateral in the figure below.

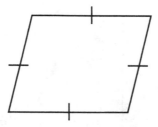

 The quadrilateral is a rhombus. A rhombus is a quadrilateral with four congruent sides. A rhombus is also a parallelogram.

 If line segments *AD* and *BC* are congruent, what type of quadrilateral is the figure below?

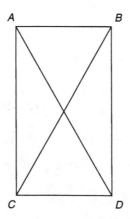

 The quadrilateral is a rectangle. A quadrilateral is a rectangle if the diagonals are congruent.

Calculate the length of the median of the trapezoid below.

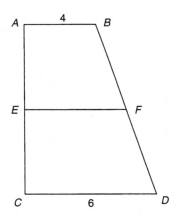

 Step 1 is to identify the median of the trapezoid.

The median is line segment *EF*.

Step 2 is to write the formula for calculating the median of the trapezoid. The formula is the length of the median equals $\frac{1}{2}$ the sum of the bases.

$$EF = \frac{1}{2}(AB + CD)$$

Step 3 is to substitute the values into the formula.

$$EF = \frac{1}{2}(4 + 6)$$

Step 4 is to calculate the length of *EF*.

$$EF = 5$$

The correct answer is the median *EF* = 5.

In this figure, a parallelogram, what is the measurement of $\angle X$? Line segment YZ is parallel to side CD.

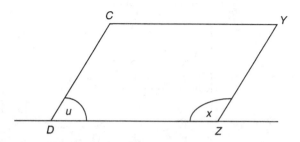

Since line segment YZ is parallel to side CD, then angle X must be equal to the supplement of angle u. Using a protractor, measure $\angle u$. Angle u is 60°. Therefore, $\angle X$ equals 180° − 60° or 120°.

12.8 Coordinate Geometry

Graph the equation $x^2 + y^2 = 16$.

The equation in this problem identifies a circle. The equation of a circle is $(x - h)^2 + (y - k)^2 = r^2$. The center of this circle is (h, k). The radius is r. Therefore, using the equation $x^2 + y^2 = 16$, the center is $(0, 0)$ and the radius is 4. See the graph in the figure below.

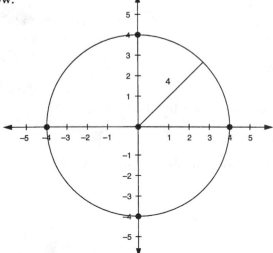

Graph the equation $x^2 + (y - 5)^2 = 25$.

 The equation in this problem identifies a circle. The equation of a circle is $(x - h)^2 + (y - k)^2 = r^2$. The center of this circle is (h, k). The radius is r. Therefore, using the equation $x^2 + (y - 5)^2 = 25$, the center is $(0, 5)$ and the radius is 5. See the graph in the figure below.

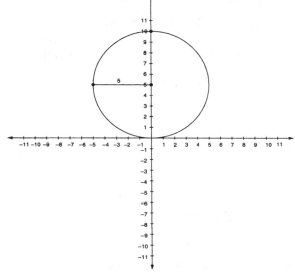

 Develop an equation of a circle with a radius of 36 and a center $(-4, -3)$.

 Step 1 is to write the equation of a circle.

$$(x - h)^2 + (y - k)^2 = r^2$$

Step 2 is to substitute the coordinates into the equation.

$$(x + 4)^2 + (y + 3)^2 = r^2$$

Step 3 is to substitute the radius into the equation.

$$(x + 4)^2 + (y + 3)^2 = 36^2$$

Calculate the distance between the points (4, 4, 4) and (−3, −4, −5).

Step 1 is to write the distance for an x,y,z plane.

$$\text{distance} = \sqrt{[(x_2 - x_1)^2 + (y_2 - y_1)^2 + (z_2 - z_1)^2]}$$

Step 2 is to substitute the values into the equation.

$$\text{distance} = \sqrt{(-3-4)^2 + (-4-4)^2 + (-5-4)^2}$$

Step 3 is to perform the operations in the parentheses.

$$\text{distance} = \sqrt{(49) + (64) + (81)}$$

Step 4 is to simplify.

$$\text{distance} = \sqrt{194}$$

The correct answer is $\sqrt{194}$.

Graph the ellipse with the equation $\dfrac{x^2}{9} + \dfrac{y^2}{4} = 1$.

Step 1 is to calculate the vertices on the x axis. To do this, set y to zero and solve for x.

$$\frac{x^2}{9} + \frac{0^2}{4} = 1$$

$$\frac{x^2}{9} + 0 = 1$$

$$x^2 = 9$$

$$x = +3, \ -3$$

Step 2 is to calculate the vertices on the y axis. To do this, set x to zero and solve for y.

$$\frac{0^2}{9} + \frac{y^2}{4} = 1$$

$$0 + \frac{y^2}{4} = 1$$

$$y^2 = 4$$

$$y = +2, \ -2$$

Step 3 is to write the vertices.

(3, 0), (–3, 0), (0, 2), and (0, –2)

Step 4 is to graph the ellipse (see the figure below).

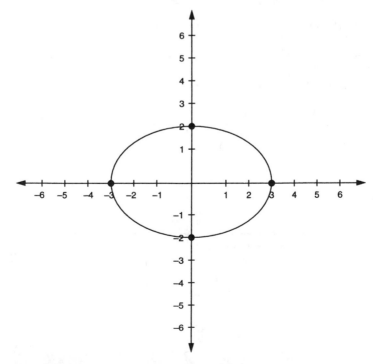

Graph the parabola with the equation $y = x^2$.

 Step 1 is to choose several values for x.

0, 1, 2, 3, -1, -2, -3

Step 2 is to substitute each value into the equation to obtain y.

$y = x^2$	$y = (0)^2$	$y = 0$
$y = x^2$	$y = (1)^2$	$y = 1$
$y = x^2$	$y = (2)^2$	$y = 4$
$y = x^2$	$y = (3)^2$	$y = 9$
$y = x^2$	$y = (-1)^2$	$y = 1$
$y = x^2$	$y = (-2)^2$	$y = 4$
$y = x^2$	$y = (-3)^2$	$y = 9$

Step 3 is to write the coordinates for the parabola.

(0, 0)   (1, 1)   (2, 4)   (3, 9)

(-1, 1) (-2, 4) (-3, 9)

Step 4 is to graph the parabola.

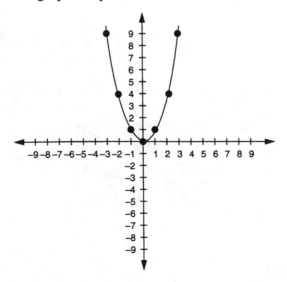

12.9   Perimeters, Areas, and Volumes

 Calculate the circumference of a circle that has a radius of 12 inches.

 Step 1 is to write the formula for the circumference of a circle.

circumference = $2\pi r$

Step 2 is to substitute the known values into the equation. Pi is approximately 3.14. The radius is 12.

circumference = $2(3.14)12$

Step 3 is to solve the equation.

circumference = 75.4

The correct answer is that the circumference = 75.4 inches.

 Calculate the area of a circle that has a diameter of 10 meters.

 Step 1 is to write the formula for the area of a circle.

$A = \pi r^2$

Step 2 is to substitute the known values into the equation. Pi is approximately 3.14. The diameter of the circle is 10 meters, so the radius is 5 meters.

$A = \pi(5)^2$

Step 3 is to solve the equation.

$A = 78.5$

The correct answer is that the area of the circle is 78.5 m².

Calculate the volume of a sphere that has a radius of 2 meters.

Step 1 is to write the formula for the volume of a sphere.

$$V = \frac{4}{3}\pi r^3$$

Step 2 is to substitute the known values into the equation. Pi is approximately 3.14 and the radius is 2 meters.

$$V = \frac{4}{3}\pi(2)^3$$

Step 3 is to solve the equation.

$V = 33.49$ or ~33.5

The correct answer is that the volume of the sphere is 33.5 m³.

Calculate the perimeter of the polygon below.

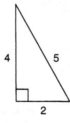

Step 1 is to write the formula for the perimeter of a triangle.

perimeter = length of side 1 + length of side 2 + length of side 3

Step 2 is to substitute the known values into the formula.

perimeter = 4 + 2 + 5

Step 3 is to solve the equation.

perimeter = 11

The correct answer is that the perimeter of the triangle is 11.

 Calculate the area of the triangle below.

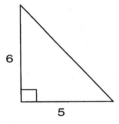

6

5

 Step 1 is to write the formula for the area of a triangle.

$$A = \frac{1}{2}bh$$

Step 2 is to substitute the known values into the formula. The base of the triangle is 5 and the height is 6.

$$A = \frac{1}{2}(5)(6)$$

Step 3 is to solve the equation.

$$A = 15$$

The correct answer is that the area of the triangle is 15.

 Calculate the volume of the cube that has a length of 5, a height of 1.5, and a width of 4.

 Step 1 is to write the formula for the volume of a cube.

$$V = L \times W \times H$$

Step 2 is to substitute the known values into the formula.

$$V = 5 \times 4 \times 1.5$$

Step 3 is to solve the equation.

$V = 30$

The correct answer is that the volume of the cube = 30.

 Calculate the volume of a pyramid that has a height of 3 feet. The area of the base of the pyramid was calculated to be 15 square feet.

 Step 1 is to write the formula for the volume of a pyramid.

$V = \dfrac{1}{3} Bh$, where B is the area of the base of the pyramid.

Step 2 is to substitute the known values into the formula.

$V = \dfrac{1}{3}(15)(3)$

Step 3 is to solve the equation.

$V = 15$

The correct answer is that the volume of the pyramid is 15 cubic feet.

 Calculate the area of the triangle below.

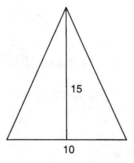

15

10

 Step 1 is to write the formula for the area of a triangle.

$A = \dfrac{1}{2} bh$

Step 2 is to substitute the known values into the formula. The base of the triangle is 10 and the height is 15.

$$A = \frac{1}{2}(10)(15)$$

Step 3 is to solve the equation.

$$A = 75$$

The correct answer is that the area of the triangle is 75.

 Calculate the volume of the cone below.

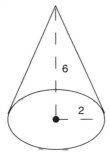

 Step 1 is to write the formula for the volume of a cone.

$$V = \frac{1}{3}\pi r^2 h$$

Step 2 is to substitute the known values into the equation.

$$V = \frac{1}{3}\pi(2)^2(6)$$

Step 3 is to solve the equation.

$$V = 25.12$$

The correct answer is that the volume of the cone is 25.12.

Calculate the volume of the cylinder below.

Step 1 is to write the formula for the volume of a cylinder.

$V = \pi r^2 h$

Step 2 is to substitute the known values into the formula.

$V = \pi (2)^2 (11)$

Step 3 is to solve the equation.

$V = 138.16$

The correct answer is that the volume is 138.16.

Calculate the total area of the cone below. Round to the nearest meter.

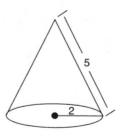

Step 1 is to write the formula for the total area of a cone.

$A = \pi r s + \pi r^2$, where s is the slant height

Step 2 is to substitute the known values into the formula.

$$A = \pi(2)(5) + \pi(2)^2$$

Step 3 is to solve the equation.

$$A = 10\pi + 4\pi = \pi(10 + 4) = \pi(14)$$

$$A = 44$$

The correct answer is 44.

 Calculate the area of the trapezoid below.

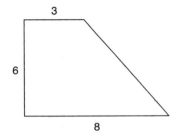

 Step 1 is to write the formula for the area of a trapezoid.

$A = \dfrac{1}{2}h(b_1 + b_2)$, where b_1 and b_2 are the bases

Step 2 is to substitute the known values into the formula.

$$A = \frac{1}{2}(6)(3 + 8)$$

Step 3 is to solve the equation.

$$A = 33$$

The correct answer is that the area of the trapezoid is 33.

Calculate the area of the rhombus below.

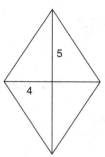

 Step 1 is to write the formula for the area of a rhombus.

$A = \frac{1}{2}(d_1)(d_2)$, where d_1 and d_2 are the diagonals of the rhombus

Step 2 is to substitute the known values into the formula.

$A = \frac{1}{2}(5)(4)$

Step 3 is to solve the equation.

$A = 10$

The correct answer is that the area of the rhombus is 10.

 Calculate the area of the parallelogram below.

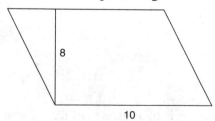

Step 1 is to write the formula for the area of a parallelogram.

$A = bh$, where b is the base and h is the height

Step 2 is to substitute the known values into the formula.

$A = 10(8)$

Step 3 is to solve the equation.

$A = 80$

The correct answer is that the area of the parallelogram is 80.

12.10 Problem Solving with Triangles

 Using the Pythagorean theorem, calculate z in the figure below.

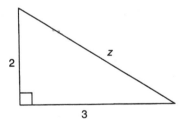

 Step 1 is to write the Pythagorean theorem.

$x^2 + y^2 = z^2$

Step 2 is to substitute the values of x and y into the equation.

$(3)^2 + (2)^2 = z^2$

Step 3 is to solve the left side of the equation.

$(3)^2 + (2)^2 = 13$

Step 4 is to rewrite the equation.

$$13 = z^2$$

Step 5 is to solve for z.

$$\sqrt{13} = z$$

The correct answer is $z = \sqrt{13}$.

 Using the Pythagorean theorem, calculate x in the figure below.

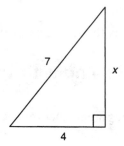

 Step 1 is to write the Pythagorean theorem.

$$x^2 + y^2 = z^2$$

Step 2 is to substitute the values of y and z into the equation.

$$x^2 + (4)^2 = 7^2$$

Step 3 is to simplify the equation.

$$x^2 + 16 = 49$$

Step 4 is to subtract 16 from both sides and rewrite the equation.

$$x^2 = 33$$

Step 5 is to solve for x.

$$x = \sqrt{33}$$

The correct answer is $x = \sqrt{33}$.

 Determine the length of side AC in the figure below.

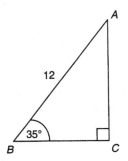

 Step 1 is to determine if sine, cosine, or tangent is needed. Since the angle and hypotenuse are known, and the opposite side needs to be determined, use sine.

Step 2 is to write the ratio for sine.

sin ∠B = opposite side/hypotenuse

In step 3, using a calculator, find the sine of ∠B.

sin ∠B = sin(35) = 0.574

Step 4 is to rewrite the ratio.

0.574 = opposite side/12

Step 5 is to multiply both sides of the equation by 12 to determine the length of the opposite side.

6.88 = length of AC

The correct answer is that the length of AC is 6.88.

Determine the length of side *AB* in the figure below.

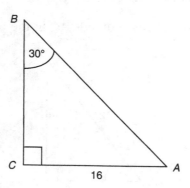

Step 1 is to determine if sine, cosine, or tangent is needed. Since the angle and opposite side are known, and the hypotenuse needs to be determined, use sine.

Step 2 is to write the ratio for sine.

sin ∠B = opposite side/hypotenuse

In step 3, using a calculator, find the sine of ∠B.

sin ∠B = sin(30) = 0.50

Step 4 is to rewrite the ratio.

0.50 = 16/hypotenuse

Step 5 is to multiply both sides of the equation by the hypotenuse.

hypotenuse(0.50) = 16

Step 6 is to divide both sides of the equation by 0.50 to determine the length of the hypotenuse.

hypotenuse = 32

The correct answer is that the length of *AB* is 32.

Determine the length of side *BC* in the figure below.

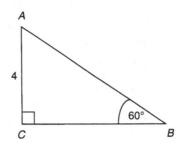

 Step 1 is to determine if sine, cosine, or tangent is needed. Since the angle and opposite side are known, and the adjacent side needs to be determined, use tangent.

Step 2 is to write the ratio for tangent.

tan ∠*B* = opposite side/adjacent side

In step 3, using a calculator, find the tangent of ∠*B*.

tan ∠*B* = tan(60) = 1.73

Step 4 is to rewrite the ratio.

1.73 = 4/adjacent side

Step 5 is to multiply both sides of the equation by the adjacent side.

1.73(adjacent side) = 4

Step 6 is to divide both sides of the equation by 1.73 to determine the length of *BC*.

2.31 = length of *BC*.

The correct answer is that the length of *BC* is 2.31.

Q A carpenter is trying to determine the height of a flagpole so she can build a new flagpole. The distance from the flagpole to the carpenter is 30 feet. The angle of inclination (the line of sight from where the carpenter is standing to the top of the flagpole) is 30°. See the figure below.

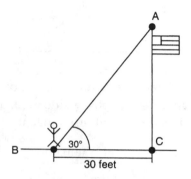

What is the height of the flagpole?

A Step 1 is to determine if sine, cosine, or tangent is needed. Since the angle and adjacent side are known, and the opposite side needs to be determined, use tangent.

Step 2 is to write the ratio for tangent.

tan ∠B = opposite side/adjacent side

In step 3, using a calculator, find the tangent of ∠B.

tan ∠B = tan(30) = 0.577

Step 4 is to rewrite the ratio.

0.577 = opposite side/30 feet

Step 5 is to multiply both sides of the equation by 30 to determine the length of AC.

17.31 = length of AC.

The correct answer is that the height of the flagpole is 17.31 feet.

Q A man is playing billiards with his friend. The cue ball hits the 7 ball at point *C*. The ball travels to point *B* and is deflected toward point *A* (the pocket). If the distance of *CB* is 14 inches, angle *C* is 60°, and ∠*B* is 90°, what is the length of *AB*? See the figure below.

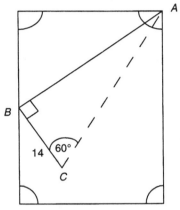

A Step 1 is to determine if sine, cosine, or tangent is needed. Since the angle and adjacent side are known, and the opposite side needs to be determined, use tangent.

Step 2 is to write the ratio for tangent.

tan ∠*C* = opposite side/adjacent side

In step 3, using a calculator, find the tangent of ∠*C*.

tan ∠*C* = tan(60) = 1.73

Step 4 is to rewrite the ratio.

1.73 = opposite side/14 inches

Step 5 is to multiply both sides of the equation by 14 to determine the length of *AB*.

24.22 = length of *AB*

The correct answer is that the distance to the pocket is 24.22 inches.

NOTES

NOTES

NOTES

NOTES

NOTES

NOTES

NOTES

NOTES

NOTES

NOTES

REA's Study Guides

Review Books, Refreshers, and Comprehensive References

Problem Solvers®

Presenting an answer to the pressing need for easy-to-understand and up-to-date study guides detailing the wide world of mathematics and science.

High School Tutors®

In-depth guides that cover the length and breadth of the science and math subjects taught in high schools nationwide.

Essentials®

An insightful series of more useful, more practical, and more informative references comprehensively covering more than 150 subjects.

Super Reviews®

Don't miss a thing! Review it all thoroughly with this series of complete subject references at an affordable price.

Interactive Flashcard Books®

Flip through these essential, interactive study aids that go far beyond ordinary flashcards.

Reference

Explore dozens of clearly written, practical guides covering a wide scope of subjects from business to engineering to languages and many more.

For our complete title list,
visit www.rea.com

Research & Education Association